RANDOM
HOUSE
LARGE
PRINT

The Mind's Eye

THE MIND'S EYE

Oliver Sacks

RANDOM HOUSE
LARGE PRINT

Published in the United States of America by
Random House Large Print in association with
Alfred A. Knopf, New York.
Distributed by Random House, Inc., New York.

Portions of this work originally appeared
in different form in **The New Yorker**.

Permission to quote previously published material
may be found following the index.

Cover design by Chip Kidd

The Library of Congress has established a Cataloging-
in-Publication record for this title.

ISBN: 978-0-7393-7803-8

www.randomhouse.com/largeprint

FIRST LARGE PRINT EDITION

Printed in the United States of America

10 9 8 7 6 5 4 3 2 1

This Large Print edition published in accord with
the standards of the N.A.V.H.

for David Abramson

Contents

Preface

I grew up in a household full of doctors and medical talk—my father and older brothers were general practitioners, and my mother was a surgeon. A lot of the dinner-table conversation was inevitably about medicine, but the talk was never just about "cases." A patient might present as a case of this or that, but in my parents' conversation, cases became biographies, stories of people's lives as they responded to illness or injury, stress or misfortune. Perhaps it was inevitable that I myself became both a physician and a storyteller.

When **The Man Who Mistook His Wife for a Hat** was published in 1985, it was given a very pleasant review by an eminent academic neurologist. The cases, he wrote, were fascinating, but he had one reservation: he thought I was being disingenuous in presenting patients as if I had come to them with no preconceptions, with little background knowledge of their conditions. Did I really read up on the scientific literature only after seeing a patient with a particular condition? Surely, he thought, I had started with a neurological theme in mind and simply sought out patients who exemplified it.

But I am not an academic neurologist, and the truth is that most practicing physicians have, apart from their broad medical education, little in-depth knowledge of many conditions, especially those which are considered rare, and thus not worthy of much time in medical school. When a patient presents himself with such a condition, we must do some research and, especially, go back to original descriptions. Typically, then, my case histories start with an encounter, a letter, a knock on the door—it is the patients' description of their experience that stimulates the more general exploration.

As a general neurologist working mostly in old-age homes, I have seen thousands of patients over the past decades. All of them have taught me something, and I enjoy seeing them—in some cases, we have been seeing each other regularly, as doctor and patient, for twenty years or more. In my clinical notes, I do my best to record what is happening with them and to reflect on their experiences. Occasionally, with the patient's permission, my notes evolve into essays.

After I began publishing case histories, starting with **Migraine** in 1970, I began receiving letters from people seeking to understand or comment on their own neurological experiences, and such correspondence has become, in a way, an extension of my practice. Thus some of the people I describe in this book are patients; others are people who have written to me after reading one of my case histories. I am grateful to all of them for agreeing to share their experiences, for such experiences enlarge the imagination and show us what is often concealed in health: the complex workings of the brain and its astounding ability to adapt and overcome disability—to say nothing of the courage and strength that individuals can show, and the inner resources they can bring to bear, in the face of neurological challenges that are almost impossible for the rest of us to imagine.

Many of my colleagues, past and present, have generously shared their time and expertise to discuss the ideas in this book or to comment on its various drafts. To all of them (and the many whom I have omitted here) I am most grateful, especially to Paul Bach-y-Rita, Jerome

Bruner, Liam Burke, John Cisne, Jennifer and John Clay, Bevil Conway, Antonio and Hanna Damasio, Orrin Devinsky, Dominic ffytche, Elkhonon Goldberg, Jane Goodall, Temple Grandin, Richard Gregory, Charles Gross, Bill Hayes, Simon Hayhoe, David Hubel, Ellen Isler at the Jewish Braille Institute, Narinder Kapur, Christof Koch, Margaret Livingstone, Ved Mehta, Ken Nakayama, Görel Kristina Näslund, Alvaro Pascual-Leone, Dale Purves, V. S. Ramachandran, Paul Romano, Israel Rosenfield, Theresa Ruggiero, Leonard Shengold, Shinsuke Shimojo, Ralph Siegel, Connie Tomaino, Bob Wasserman, and Jeannette Wilkens.

I could not have completed this book without the moral and financial support of a number of institutions and individuals, and I am enormously indebted to them, above all to Susie and David Sainsbury, Columbia University, **The New York Review of Books, The New Yorker,** the Wylie Agency, the MacDowell Colony, Blue Mountain Center, and the Alfred P. Sloan Foundation. I am grateful, too, to the many people at Alfred A. Knopf, Picador UK, Vintage Books, and my other publishers around the world.

Several correspondents have contributed ideas or descriptions to this book, including

Joseph Bennish, Joan C., Larry Eickstaedt, Anne F., Stephen Fox, J. T. Fraser, and Alexandra Lynch.

I am grateful to John Bennet at **The New Yorker** and Dan Frank at Knopf, superb editors who have improved this book in many ways; and to Allen Furbeck for his help with the illustrations. Hailey Wojcik typed many of the drafts and contributed research and virtually every other type of assistance, to say nothing of deciphering and transcribing the almost 90,000 words of my "melanoma journals." Kate Edgar has, for the past twenty-five years, filled a unique role as collaborator, friend, editor, organizer, and much else. She has incited me, as always, to think and write, to see from different perspectives, but always to return to the center.

Above all, I am indebted to my subjects or patients and their families: Lari Abraham, Sue Barry, Lester C., Howard Engel, Claude and Pamela Frank, Arlene Gordon, Patricia and Dana Hodkin, John Hull, Lilian Kallir, Charles Scribner, Jr., Dennis Shulman, Sabriye Tenberken, and Zoltan Torey. They have not only allowed me to write about their experiences and quote their descriptions; they have commented on drafts, introduced me to other people and resources, and, in many cases, become good friends.

Finally, I must express my deepest gratitude to my physician, David Abramson; to him I dedicate this book.

O.W.S.

New York
June 2010

The Mind's Eye

Sight Reading

In January of 1999, I received the following letter:

> Dear Dr. Sacks,
>
> My (very unusual) problem, in one sentence, and in non-medical terms, is: I can't read. I can't read music, or anything else. In the ophthalmologist's office, I can read the individual letters on the eye chart down to the last line. But I cannot read words, and music gives me the same problem. I have struggled with this for years, have been to the best doctors, and no one has been able to help. I would be ever so happy and grateful if you could find the time to see me.
>
> Sincerely yours,
> Lilian Kallir

I phoned Mrs. Kallir—this seemed to be the thing to do, although I normally would have written back—because although she apparently had no difficulty writing a letter, she had said that she could not read at all. I spoke to her and arranged to see her at the neurology clinic where I worked.

Mrs. Kallir came to the clinic soon afterward—a cultivated, vivacious sixty-seven-year-old woman with a strong Prague accent—and related her story to me in much more detail. She was a pianist, she said; indeed, I knew her by name, as a brilliant interpreter of Chopin and Mozart (she had given her first public concert at the age of four, and Gary Graffman, the celebrated pianist, called her "one of the most naturally musical people I've ever known").

The first intimation of anything wrong, she said, had come during a concert in 1991. She was performing Mozart piano concertos, and there was a last-minute change in the program, from the Nineteenth Piano Concerto to the Twenty-first. But when she flipped open the score of the Twenty-first, she found it, to her bewilderment, completely unintelligible. Although she saw the staves, the lines, the individual notes sharp and clear, none of it seemed to hang together, to make sense. She thought the difficulty must have something to do with her eyes. But she went on to perform the concerto flawlessly from memory, and dismissed the strange incident as "one of those things."

Several months later, the problem recurred, and her ability to read musical scores began to fluctuate. If she was tired or ill, she could hardly read them at all, though when she was fresh, her sight-reading was as swift and easy as ever. But in general the problem worsened, and though she continued to teach, to record, and to give concerts around the world, she depended increasingly on her musical memory and her extensive repertoire, since it was now becoming impossible for her to learn new music by sight. "I used to be a fantastic sight reader," she said, "easily able to play a Mozart concerto by sight, and now I can't."

Occasionally at concerts she experienced lapses of memory, though Lilian (as she asked me to call her) was adept at improvising and could usually cover these. When she was at ease, with friends or students, her playing seemed as good as ever. So, through inertia, or fear, or a sort of adjustment, it was possible for her to overlook her peculiar problems in reading music, for she had no other visual problems, and her memory and ingenuity still allowed her a full musical life.

In 1994, three years or so after she had

first noticed problems reading music, Lilian started to have problems with reading words. Here again, there were good days and bad, and even times when her ability to read seemed to change from moment to moment: a sentence would look strange, unintelligible at first; then suddenly it would look fine, and she would have no difficulty reading it. Her ability to write, however, was quite unaffected, and she continued to maintain a large correspondence with former students and colleagues scattered throughout the world, though she depended increasingly on her husband to read the letters she received, and even to reread her own.

Pure alexia, unaccompanied by any difficulty in writing ("alexia sine agraphia") is not that uncommon, although it usually comes on suddenly, following a stroke or other brain injury. Less often, alexia develops gradually, as a consequence of a degenerative disease such as Alzheimer's. But Lilian was the first person I had encountered whose alexia manifested first with musical notation, a musical alexia.

By 1995 Lilian was beginning to develop additional visual problems. She noticed that she tended to "miss" objects to the right, and, after some minor mishaps, she decided that she had best give up driving.

She had sometimes wondered whether her

strange problem with reading might be neurological rather than ophthalmological in origin. "How can I recognize individual letters, even the tiny ones on the bottom line of the eye doctor's chart, and yet be unable to read?" she wondered. Then, in 1996, she started to make occasional embarrassing mistakes, such as failing to recognize old friends, and she found herself thinking of a case history of mine she had read years before, entitled "The Man Who Mistook His Wife for a Hat," about a man who could see everything clearly but recognize nothing. She had chuckled when she had first read it, but now she started to wonder whether her own difficulties might be eerily similar in nature.

Finally, five years or more after her original symptoms, she was referred to a university neurology department for a full workup. Given a battery of neuropsychological tests—tests of visual perception, of memory, of verbal fluency, etc.—Lilian did particularly badly in the recognition of drawings: she called a violin a banjo, a glove a statue, a razor a pen, and pliers a banana. (Asked to write a sentence, she wrote, "This is ridiculous.") She had a fluctuating lack of awareness, or "inattention," to the right, and very poor facial recognition (measured by recognition of photographs of famous public figures). She could read, but only slowly, letter by letter. She would read a "C," an "A," a "T," and then, laboriously, "cat," without recognizing the word as a whole. Yet if she was shown words too quickly to decipher in this way, she could sometimes correctly sort them into general categories, such as "living" or "nonliving," even though she had no conscious idea of their meaning.

In contrast to these severe visual problems, her speech comprehension, repetition, and verbal fluency were all normal. An MRI of her brain was also normal, but when a PET scan was performed—this can detect slight changes in the metabolism of different brain areas, even when they appear anatomically normal—Lilian was found to have diminished metabolic activity in the posterior part of the brain, the visual cortex. This was more marked on the left side. Noting the gradual spread of difficulties in visual recognition—first of music, then of words, then of faces and objects—her neurologists felt she must have a degenerative condition, at present confined to the posterior parts of the brain. This would probably continue to worsen, though very slowly.

The underlying disease was not treatable in any radical sense, but her neurologists suggested that she might benefit from certain strategies: "guessing" words, for example, even when she could not read them in the ordinary way (for it was clear that she still possessed some mechanism that allowed unconscious or preconscious recognition of words). And they suggested that she might also use a deliberate, hyperconscious inspection of objects and faces, making particular note of their distinctive features, so that these could be identified in future encounters, even if her normal "automatic" powers of recognition were impaired.

In the three years or so that had elapsed between this neurological exam and her first visit to me, Lilian told me, she had continued to perform, though not as well, and not as frequently. She found her repertoire diminishing, because she could no longer check even familiar scores by vision. "My memory was no longer fed," she remarked. Fed visually, she meant—for she felt that her auditory memory, her auditory orientation, had increased, so that she could now, to a much larger degree than before, learn and reproduce a piece by ear. She could not only play a piece in this way (sometimes after only a single hearing); she could rearrange it in her mind. Nonetheless, there was, on balance, a shrinkage of her repertoire, and she began to avoid giving public concerts. She continued to play in more informal settings and to teach master classes at the music school.

Handing me the neurological report from

1996, she commented, "The doctors all say, 'Posterior cortical atrophy of the left hemisphere, very atypical,' and then they smile apologetically—but there's nothing they can do."

When I examined Lilian, I found that she had no problem matching colors or shapes, or recognizing movement or depth. But she showed gross problems in other areas. She was unable now to recognize individual letters or numerals (even though she still had no difficulty writing complete sentences). She had, too, a more general visual agnosia, and when I presented her with pictures to identify, it was difficult for her even to recognize pictures **as** pictures—she would sometimes look at a column of print or a white margin, thinking it was the picture I was quizzing her about. Of one such picture, she said, "I see a V, very elegant—two little dots here, then an oval, with little white dots in between. I don't know what it's supposed to be." When I told her it was a helicopter, she laughed, embarrassed. (The V was a sling; the helicopter was unloading food supplies for refugees. The two little dots were wheels, the oval the helicopter's body.) Thus she was now seeing only individual features of an object or picture, failing to synthesize them, to see them as a whole, much less to interpret them correctly. Shown a photograph of a face, she could perceive that the person was wearing glasses, nothing else. When I asked if she could see clearly, she said, "It's not a blur, it's a mush"—a mush consisting of clear, fine, sharp but unintelligible shapes and details.

Looking at the drawings in a standard neurological test booklet, she said of a pencil, "Could be so many things. Could be a violin . . . a pen." A house, however, she immediately recognized. Regarding a whistle, she said, "I have no idea." Shown a drawing of scissors, she looked steadfastly at the wrong place, at the white paper below the drawing. Was Lilian's difficulty in recognizing drawings due simply to their "sketchiness," their two-dimensionality, their poverty of information? Or did it reflect a higher-order difficulty with the perception of representation as such? Would she do better with real objects?

When I asked Lilian how she felt about herself and her situation, she said, "I think I am dealing with it very well, most of the time . . . knowing it is not getting better, but only slowly worse. I've stopped seeing neurologists. I always hear the same thing. . . . But I am a very resilient person. I don't tell my friends. I don't want to burden them, and my little story is not very promising. A dead end. . . . I have a good sense of humor. And that's it, in a nutshell. It is depressing, when I think of it—frustrations daily. But I have many good days and years ahead."

After Lilian left, I was unable to find my medical bag—a black bag with some similarities (I now remembered) to one of the several bags she had brought. Going home in the taxi, she realized that she had taken the wrong bag when she saw a red-tipped object sticking out of it (my long, red-tipped reflex hammer). It had attracted her attention, by its color and shape, when she saw it on my desk, and now she realized her mistake. Returning, breathless and apologetic, to the clinic, she said, "I am the woman who mistook the doctor's bag for her handbag."

Lilian had done so badly on the formal tests of visual recognition that I had difficulty imagining how she managed in daily life. How did she recognize a taxi, for example? How could she recognize her own home? How could she shop, as she told me she did, or recognize foods and serve them on a table? All this and much more—an active social life, traveling, going to concerts, and teaching—she did by herself when her husband, who was also a musician, went to Europe for weeks at a time. I could get no idea of how she accomplished this from seeing her dismal performance in the artificial, impoverished atmosphere of a neurology clinic. I had to see her in her own familiar surroundings.

The following month, I visited Lilian at home, home being a pleasant apartment in upper Manhattan where she and her husband had lived for more than forty years. Claude was a charming, genial man about the same age as his wife. They had met as music students at Tanglewood nearly fifty years earlier and had pursued their musical careers in tandem, often performing onstage together. The apartment had a friendly, cultured atmosphere, with a grand piano, a great many books, photographs of their daughter and of friends and family, abstract modernist paintings on the wall, and mementos of their trips on every available surface. It was crowded—rich in personal history and significance, I imagined, but a nightmare, a complete chaos, for someone with visual agnosia. This, at least, was my first thought as I entered, negotiating my way between tables full of knickknacks. But Lilian had no difficulty with the clutter and threaded her way confidently through the obstacles.

Since she had had such difficulty on the drawing-recognition test, I had brought a number of solid objects with me, wondering if she would do better with these. I started with some fruit and vegetables I had just bought, and here Lilian did surprisingly well. She instantly identified "a beautiful red pepper," recognizing it from across the room; a banana, too. She was momentarily uncertain whether the third object was an apple or a tomato, though she soon decided, correctly, on the former. When I showed her a small plastic model of a wolf (I keep a variety of such objects, for perceptual testing, in my medical bag), she exclaimed, "A marvelous animal! A baby elephant, perhaps?" When I asked her to look more closely, she decided it was "a kind of dog."

Lilian's relative success in naming solid objects, as opposed to drawings of them, again made me wonder whether she had a specific agnosia for representations. The recognition of representations may require a sort of learning, the grasping of a code or convention, beyond that needed for the recognition of objects. Thus, it is said, people from primitive cultures who have never been exposed to photographs may fail to recognize that they are representations of something else. If a complex system for the recognition of visual representations must be specially constructed by the brain, this ability might be lost through damage to that system by a stroke or disease, just as the learned understanding of writing, say, or any other acquired ability may be lost.

I followed Lilian into the kitchen, where she set about taking the kettle off the stove and pouring boiling water into the teapot. She seemed to navigate her crowded kitchen well, knowing, for instance, that all the skillets and pots were hung on hooks on one wall, various supplies kept in their regular places. When we opened the refrigerator and I quizzed her on the contents, she said, "O.J., milk, butter on the top shelf—and a nice sausage, if you're interested, one of those Austrian things . . . cheeses." She recognized the eggs in the fridge door and, when I asked her, counted them correctly, moving her finger from egg to egg as she did so. I could see at a glance that there were eight—two rows of four—but Lilian, I suspect, could not perceive the eightness, the gestalt, easily and had to enumerate the eggs one by one. And the spices, she said, were "a disaster." They all came in identical red-topped bottles, and, of course, she could not read the labels. So: "I smell them! . . . And I call for help some of the time." With the microwave oven, which she used often, she said, "I don't see the numbers. I do it by feel—cook, try, see if it needs a bit more."

Though Lilian could scarcely recognize anything in the kitchen visually, she had organized it in such a way that mistakes rarely, if ever, occurred, utilizing a sort of informal classification system instead of a direct perceptual gnosis. Things were categorized not by meaning but by color, by size and shape, by position, by context, by association, somewhat as an illiterate person might arrange the books in a library. Everything had its place, and she had memorized this.

Seeing how Lilian inferred the character of the objects around her in this way, using color, above all, as a marker, I wondered how she would do with objects of similar appearance, like the fish knives and the steak knives, which looked almost the same. This was a problem, she confessed, and she often confused them. Perhaps, I suggested, she could use an artificial marker, a little green dot for the fish knives, a red one for the steak knives, so that she could see the difference at a glance. Lilian said she had already thought about this but was not sure she wanted to "flaunt" her problem to others. What would her guests think of color-coded cutlery and dishes, or a color-coded apartment? ("Like a psychological experiment," she said, "or an office.") The "unnaturalness" of such an idea disturbed her, but if the agnosia got worse, she agreed, she might need it.

In some cases where Lilian's categorization

system did not work, such as using the microwave, she could operate by trial and error. But if objects were not in their place, major difficulties could appear. This showed itself startlingly at the end of my visit. The three of us—Lilian, Claude, and I—had sat down at the dining room table. Lilian had laid the table, put out biscotti and cakes, and now brought in a steaming pot of tea. She chatted as we ate, but retained a certain watchfulness, monitoring the position and movement of every dish, tracking everything (I later realized), so that it did not get "lost." She got up to take the empty dishes into the kitchen, leaving only the biscotti, which she saw that I especially liked. Claude and I chatted for a few minutes—our first talk alone—pushing the plate of biscotti between us.

When Lilian came back, and I packed my bag and prepared to go, she said, "You must take the rest of the biscotti with you"—but now, bizarrely, she could not find them, and became upset, almost frantic, at this. They were right on the table in their dish, but since the dish had been moved she no longer knew where they were, or even where to look. She seemed to have no strategy for looking. She was, however, quite startled to see my umbrella on the table. She failed to recognize it as an umbrella, noticing only that something curved and twisted had appeared—and wondered, for a half-serious moment, if it was a snake.

Before I left, I asked Lilian to go to the piano, asked if she would play something for me. She hesitated. It was clear that she had lost a good deal of her confidence. She started beautifully, on a Bach fugue, but broke off, apologetically, after a few bars. Seeing a volume of Chopin mazurkas on the piano, I asked about those, and, encouraged, she closed her eyes and played two of the Opus 50 mazurkas without faltering, and with great brio and feeling.

She told me afterward that the printed music was just "lying around," saying, "It throws me off to see the score, people turning pages, my hands, or the keyboard," and that, in such circumstances, she might make mistakes, especially with her right hand. She had to close her eyes and perform nonvisually, using only her "muscle memory," and her fine ear.

What could I say about the nature and progress of Lilian's strange disease? It had clearly advanced somewhat since her neurological examination three years before, and there were hints—though no more than hints—that her problems might no longer be purely visual. In particular, she occasionally had difficulty naming objects even when she recognized them, and would speak of a "thingmy" when she could not get the word.

I had ordered a new MRI to compare with her earlier one, and it showed that there was now some shrinkage of the visual areas on both sides of the brain. Was there any sign of real damage elsewhere? It was difficult to tell, although I suspected that there might have been some shrinkage in the hippocampi, too—parts of the brain crucial for the registration of new memories. But the damage was still largely confined to the occipital and occipitotemporal cortex, and it was clear that the rate of advance was very slow.

When I discussed these MRI findings with Claude, he stressed that in speaking with Lilian I should avoid certain terms, above all the frightening label of Alzheimer's disease. "It's not Alzheimer's disease, is it?" he said. Clearly, this had been much on their minds.

"I'm not sure," I said. "Not in the ordinary sense. One should see it as something rarer—and more benign."

Posterior cortical atrophy, PCA, was first formally described by Frank Benson and his colleagues in 1988, although it has undoubtedly existed, unrecognized, for much longer.

But Benson et al.'s paper provoked a rush of recognition, and dozens of cases have now been described.

People with PCA preserve elementary aspects of visual perception, such as acuity or the ability to detect movement or color. But they tend to experience complex visual disturbances—difficulties reading or recognizing faces and objects, occasionally even hallucinations. Their visual disorientation may become profound: some patients get lost in their own neighborhoods or even in their own homes; Benson called this "environmental agnosia." Other difficulties commonly follow: left-right confusion, difficulty in writing and calculation, even an agnosia for one's own fingers, a tetrad of problems sometimes called Gerstmann's syndrome. Sometimes patients with PCA may be able to recognize and match colors but unable to name them, a so-called color anomia. More rarely, there can be a difficulty in visual targeting and tracking movements.

In contrast to these difficulties, memory, intelligence, insight, and personality tend to be preserved until late in the course of the disease. Every patient described by Benson, he writes, "could present his or her own history, was aware of current events, and showed considerable insight into his or her predicament."

Although PCA is clearly a degenerative brain

disease, it seems quite different in character from the commoner forms of Alzheimer's, where gross changes in memory and thinking, in the comprehension and use of language, and often in behavior and personality tend to occur, and insight into what is happening (perhaps mercifully) is generally lost early on.

In Lilian's case, the course of the disease seemed to have been relatively benign, for even nine years after her first symptoms, she did not get lost in her own home or neighborhood.

I could not help making a comparison, as Lilian herself had, with my patient Dr. P., "the man who mistook his wife for a hat." Both of them were highly gifted professional musicians; both developed severe visual agnosias, while remaining remarkably intact in many other ways; and both had discovered or developed ingenious ways around their problems, so that it was possible for them to keep teaching at the highest level in music colleges, despite what might appear to be quite devastating disabilities.

The actual ways in which Lilian and Dr. P. coped with their illnesses were very different, though—a reflection in part of the severity of their symptoms, and in part of differences in temperament and training. Dr. P. was already in grave trouble when I saw him, barely three years after his initial symptoms. He had not only visual difficulties but tactile ones, too—he grasped his wife's head and mistook it for a hat. He showed a sort of levity or indifference, and little insight into the fact that he was ill, and he often confabulated to make up for the fact that he could not identify what he was seeing. This was in strong contrast to Lilian, who, nine years after her first symptoms, had no substantial problems outside her visual ones, was still able to travel and teach, and showed acute insight into her own condition.

Lilian could still identify objects by inference, using her intact perception of color, shape, texture, and movement, along with her memory and intelligence. Dr. P. could not. He could not, for instance, identify a glove by sight or by feel (despite being able to describe it in almost absurdly abstract terms, as "a continuous surface infolded on itself [with] five outpouchings, if this is the word . . . a container of some sort?")—until, by accident, he got it onto his hand. He was, in general, almost wholly dependent on **doing** things, on action, on flow. And singing, which for him was the most natural, irrepressible activity in the world, allowed him to bypass his agnosia to some extent. He had all sorts of songs that he would hum or sing: dressing songs, shaving songs, action songs. Music, he had found, could organize his activities, his daily life.[1] This was not the case with Lilian. Her great musicality was also preserved, but it did not play a comparable role in her daily life; it was not, for her, a strategy for dealing with agnosia.

A few months later, in June of 1999, I again visited Lilian and Claude in their apartment—Claude was just back from his weeks in Europe, and Lilian, I gathered, had been mov-

---

1. I saw Dr. P. in 1978, ten years before Benson and his colleagues described PCA. I was puzzled by the picture Dr. P. presented, the paradoxes of his illness. Clearly, he had a degenerative brain disease, yet it seemed quite different from any form of Alzheimer's disease I had seen. But if not Alzheimer's, then what did he have? When I read about PCA in 1988—Dr. P. had died in the meantime—I wondered whether this could have been his diagnosis.

PCA, however, is only an anatomical diagnosis; it denotes the part of the brain affected most but says nothing of the underlying disease process, nothing of why these parts of the brain are damaged.

When Benson described PCA, he had no information regarding its underlying pathology. His patients might have Alzheimer's disease, he thought, but if so, it was Alzheimer's with a strikingly atypical presentation. They might have Pick's disease, a degenerative brain disorder more commonly affecting the frontal and temporal lobes of the brain. They might even, Benson speculated, have vascular rather than degenerative disease, an accumulation of small blockages in the watershed zone between the posterior and carotid circulations of the brain.

ing freely within a four-block radius of their apartment, going to her favorite restaurant, shopping, doing errands. When I arrived, I saw that Lilian had been sending cards to her friends all over the world—there were envelopes addressed to Korea, to Germany, to Australia, to Brazil, scattered all over the table. Her alexia, clearly, had not diminished her correspondence, though the names and addresses sometimes straggled over the envelope. She seemed to be managing well in her own apartment, but how did she deal with shopping and the challenges of a busy New York neighborhood, even her own?

"Let's go out, let's wander," I said. Lilian immediately started singing "**Der Wanderer**"—she loves Schubert—and then the elaboration of this in the **Wanderer** Fantasy.

In the elevator, she was greeted by some neighbors. It was not clear to me whether she recognized them visually or by their voices. She instantly recognized voices, sounds of all sorts; indeed, she seemed hyperattentive here, as she was to colors and shapes. They had assumed a special importance as cues.

She had no difficulty crossing the street. She could not read the "Walk" and "Don't Walk" signs, but she knew their relative position and color; knew, too, that she could walk when the sign was blinking. She pointed out a synagogue on the corner opposite; other shops she identified by shapes or colors, as with her favorite diner, which had alternating black and white tiles.

We went into a supermarket and got a cart—she headed instantly to the alcove where these were. She had no difficulty in finding the fruit and vegetable section, or in identifying apples, pears, carrots, yellow peppers, and asparagus. She could not at first name a leek but said, "Is it a cousin of an onion?" and then got the missing word, "leek." She was puzzled by a kiwifruit, until I let her handle it. (She thought it "delightfully furry, like a little mouse.") I reached up for an object hanging above the fruit. "What is this?" I asked. Lilian squinted, hesitated. "Is it edible? Paper?" When I let her touch it, she burst into somewhat embarrassed laughter. "It's an oven glove, a pot holder," she said. "How could I be so silly?"

When we moved to the next section, Lilian called out, "Salad dressings on the left, oils on the right," in the manner of a department-store elevator operator. She had obviously mapped the entire supermarket in her head. Wanting a particular tomato sauce, one of a dozen different brands, she picked it out because it had "a deep-blue rectangle and below that a yellow circle" on its label. "Color is of the essence," she emphasized again. This is her most immediately visible cue, recognizable when nothing else is. (For that reason, fearing we might be separated, I had dressed entirely in red for our visit, knowing that it would allow her to spot me instantly if we did.)

But color was not always enough. If confronted with a plastic container, she might have no idea whether it contained peanut butter or cantaloupe. Often, she found that the simplest strategy was to bring in a used can or carton and ask someone for help in matching it.

As we left the market, she accidentally crashed the shopping cart into a pile of shopping baskets to her right. Such accidents, when they happen, are always to the right, because of her impaired visual awareness to this side.

Some months later, I arranged to see Lilian in my own office rather than at the clinic, where she had come before. She arrived promptly, having made her way to Greenwich Village from Penn Station. She had been in New Haven the night before, where her husband had given a concert, and he had seen her onto a train that morning. "I know Penn Station like the back of my hand," she said, so she did not have problems there. But outside, in the melee of people and traffic, she noted, "there were many moments when I had to ask." When I inquired about how she had been doing, she said her agnosia was getting worse. "When you and I went to the market together, there were many things I could recognize easily. Now, if I want to buy the same things, I have to ask people." In general, she had to ask others to identify objects for her, or to help her if there were awkward steps, sudden changes of level, or irregularities in the ground. She depended more on touch and on hearing (to make sure, for instance, that she was facing the right way). And she depended increasingly on her memory, her thinking, her logic and common sense to help negotiate what would otherwise be—visually—an unintelligible world.

Yet, in my office, she immediately recognized a picture of herself on a CD cover, playing Chopin. "It looks slightly familiar," she said with a smile.

I asked her what she saw on a certain wall of my office. First, she turned her chair not to the wall but to the window, and said, "I see buildings." Then I rotated her chair for her until she faced the wall. I had to take her through it bit by bit. "Do you see lights?" Yes, there, and there. It took a little while to establish that she was looking at a sofa beneath the lights, though its color was commented on at once. She observed something green lying on the sofa, and astonished me by saying, correctly, that it was a stretch cord. She said she had been given such a cord by her physiotherapist. Asked what she saw above the couch (a painting with abstract geometric forms), she said, "I see yellow . . . and black." What is it? I asked. Something to do with the ceiling, Lilian hazarded. Or a fan. A clock. Then she added, "I haven't really found out whether it is one item or many." It was in fact a painting done by another patient, a colorblind painter. But clearly Lilian had no idea that it was a painting, was not even sure that it was a single object, and thought that it might be part of the structure of the room.

I found all this puzzling. How was it that she could not clearly distinguish a striking painting from the wall itself, yet could instantly recognize a small photograph of herself on a CD? How could she identify a slender green stretch cord while failing to see, or recognize, the sofa it was on? And there had been innumerable such inconsistencies before.

I wondered how she could read the time, since she was wearing a wristwatch. She could not read the numbers, she said, but could judge the position of the hands. I then showed her, mischievously, a strange clock I have, in which the numbers are replaced by the symbols of elements (H, He, Li, Be, etc.). She did not perceive anything the matter with this, since for her the chemical abbreviations were no more or less unintelligible than numerals would have been.

We went out for a walk, I in a bright-colored hat for recognition. Lilian was bewildered by the objects in one shop window—but so was I. This was a Tibetan-handicrafts shop, but they could have been Martian handicrafts, given the exotic unfamiliarity of everything. The shop next to this one, curiously, she recognized at once, and mentioned having passed it on her way to my office. It was a clock shop, with dozens of clocks of different sizes and shapes. She told me later that her father had had a passion for clocks.

A padlock on the door of another store was a total puzzle, though Lilian thought it might be something "to open up . . . like a hydrant." The moment she touched it, though, she knew what it was.

We stopped briefly for coffee; then I took her to my apartment, on the next block. I wanted her to try my grand piano, an 1894 Bechstein. Entering my apartment, she immediately identified the grandfather clock in the hall. (Dr. P., by contrast, had tried to shake hands with a grandfather clock.)

She sat at the piano and played a piece—a piece that I found puzzling, for it seemed familiar to me in a way, yet unfamiliar, too. Lilian explained that it was a Haydn quartet she had heard on the radio and been enchanted by a couple of years before and which she had longed to play herself. So she had arranged it for the piano, and had done so entirely in her head, overnight. She had occasionally arranged pieces for the piano before her alexia, using manuscript paper and the original score, but when this became impossible, she found that she could do it wholly by ear. She felt that her musical memory, her musical imagery, had become stronger, more tenacious, but also more flexible, so that she could hold the most complex music in her mind, then rearrange it and replay it mentally, in a way that would have been impossible before. Her continually strengthening powers of musical memory and imagery had become crucial to her, kept her going since the onset of her visual difficulties, nine years earlier.[2]

---

2. I was reminded, when Lilian told me this, of a patient I had seen in the hospital some years before, who had overnight become totally paralyzed from a spinal cord infection, a fulminating myelitis. When it became evident that no recovery was forthcoming, she fell into despair, felt that her life was over—not only the great things of life but the little familiar pleasures of each day, like doing the **New York Times** crossword, to which she was addicted. She requested that the **Times**

Lilian's obvious confusion about what was what in my office, and in the little streets and shops around it, brought home to me how dependent she was on the familiar, the memorized; how anchored she was to her own apartment and her own neighborhood. In time, perhaps, if she were to visit a place frequently, she would gradually become more familiar with it, but this would be a hugely complex enterprise, demanding great patience and resourcefulness, a whole new system of categorization and memorization. It was clear to me, after this one visit of Lilian's to my office, that in the future I should stick to house calls, visiting her in her own apartment, where she felt organized, in control, at home. Going out, for her, was becoming an increasingly surreal visual challenge, full of fantastic and sometimes frightening misperceptions.

---

be brought to her each day, so that at least she could look at the puzzle, get its configuration, run her eyes along the clues. But when she did this something extraordinary happened, for as she looked at the clues, the answers seemed to write themselves in their spaces. Her visual imagery strengthened over the next few weeks, until she found that she was able to hold the entire crossword and its clues in her mind after a single, intense inspection, and then solve it, mentally, at her leisure later in the day. This became a source of great solace to her, in her paralysis; she had had no idea, she later told me, that such powers of memory and imagery were available to her.

. . .

Lilian wrote to me again in August of 2001, expressing growing concern. She said she hoped I might be able to come soon for a visit, and I suggested the following weekend.

She stood by her door to welcome me, knowing, as she did, my own (lifelong) defects of visual and topographic recognition, my confusion of left and right, and my inability to find my way around inside buildings. She welcomed me with great warmth, but also a touch of anxiety, which seemed to hover throughout the visit.

"Life is difficult," she began, after she had seated me and given me a glass of seltzer. She had trouble finding the seltzer in her refrigerator, and, not seeing the bottle, which was "hidden" behind a jug of orange juice, she had taken to exploring the refrigerator by hand, groping for a bottle of the right shape. "It is not getting better. . . . The eyes are very bad." (She knows, of course, that her eyes are fine, and that it is the visual parts of the brain that are declining—indeed, she realized this before anyone else—but she finds it easier, more natural, to refer to her "bad eyes.") When I had gone shopping with her two years before, she had seemed to recognize almost everything she saw, or at least had it coded by shape and color and location, so that she hardly ever needed help. At that time, too, she moved infallibly about her kitchen, never losing anything, working efficiently. Today, she "lost" both the seltzer and the schmaltz herrings—a losing that entailed not only forgetting where she had put them but not recognizing them when she saw them. I observed that the kitchen was less organized than it had been before—and organization is crucial in her situation.

Lilian's anomia, her problems with finding words, had increased, too. When I showed her some kitchen matches, she recognized them at once, visually, but could not say the word "match," saying, instead, "That is to make fire." The Sweet'n Low, similarly, she could not name, but identified as "Better Than Sugar." She was well aware of these difficulties, and of her strategies for dealing with them. "When I can't say something," she explained, "I circumscribe."

She said that although she had recently traveled to Ontario, to Colorado, and to Connecticut with her husband, she would not have been able to do this by herself, as she had only a few years before. She felt that she remained quite capable of looking after herself at home when Claude was away. Still, she said, "When I am alone, it is lousy. I'm not complaining—I'm describing."

While Lilian was in the kitchen at one point, I asked Claude how he felt about these problems. He expressed sympathy and understanding, but added, "My impatience is provoked sometimes when I think that some of her weaknesses may be exaggerated. I'll give you an example. I get puzzled, annoyed sometimes, because Lilian's 'blindness' is sometimes 'selective.' Last Friday, she noticed that a painting was hung lopsidedly by a few millimeters. And sometimes she comments on people's facial expressions in tiny photographs. She will touch a spoon and ask, 'What is this?' and then five minutes later look at a vase and say, 'We have a similar one.' I have found no pattern, only inconsistency. What should my attitude be when she grabs a cup and says, 'What's this?' I sometimes don't tell her. But this may be wrong, and the effect disastrous. What should I say?"

This was, indeed, a very delicate matter. How much should he intervene when she was faced with perceptual bewilderment? How much should we prompt a friend or a patient when he has forgotten someone's name? How much do I myself—with no sense of direction—wish to be saved from blundering off in the wrong direction or left to battle out the right way by myself? How much do any of us like to be "told" anything? The question was especially vexing with Lilian, for, while she needed to work things out, fend for herself, her visual difficulties were becoming more severe all the time, and they sometimes threatened, as Claude observed, to throw her into a panic of disorientation. I could suggest no rule, I said to Claude, except that of tact: each situation would call for its own solution.

But I, too, was puzzled by the extraordinary variations in Lilian's visual function. Some of them, it seemed, went with the reduced and unstable function of her damaged visual cortex—just as, ten years earlier, when the first problems appeared, her ability to read music would come and go. Some of the variations, I thought, might reflect fluctuations in blood flow. But some of the variations seemed to go with a decreasing ability, for whatever reason, to compensate in her usual way. Her ability to make use of her memory and her intellectual powers in place of direct visual recognition, I now felt, might also be diminishing at this point. Thus it was more important than ever for Lilian to "code" things, to provide easily used sensory clues—above all, color, to which she remained intensely sensitive.

What intrigued me especially was Claude's

mention of Lilian's sudden abilities—her ability, for example, to perceive facial expressions on a tiny photograph, even though most of the time she had difficulty recognizing people at all. I could not help wondering whether this was an example of the preconscious abilities she had shown on earlier testing—as when she could categorize words, even though she could not recognize the objects they represented, as "living" or "nonliving." Such unconscious recognition might be possible to some extent despite her agnosia, despite her cortical damage, because it made use of other, still intact mechanisms in the visual system.

An extraordinary firsthand account of "musical alexia with recovery" was published by Ian McDonald in 2006. It was the first such personal account to be published, and was doubly remarkable because McDonald himself was both a neurologist and a fine amateur musician. His musical alexia (along with other problems, including difficulties with calculation, face-blindness, and topographic disorientation) was caused by an embolic stroke, and he was to make a complete recovery.[3] He stressed that,

3. McDonald also lost, temporarily, the ability to play the piano accurately and expressively, a problem Lilian did not have.

even though there was gradual improvement in his ability to read music, especially associated with practice, his musical alexia fluctuated considerably from day to day.

Lilian's physicians initially thought that she, too, had had a stroke and that the variations in her abilities might go with this. But such fluctuations are typical of any neural system that has sustained damage, irrespective of the cause. Patients with sciatica from nerve-root compression have good and bad days, as do patients with impairments of sight or hearing. There is less reserve, less redundancy, when a system is damaged, and it is more easily thrown off by adventitious factors such as fatigue, stress, medications, or infections. Such damaged systems are also prone to spontaneous fluctuations, as my **Awakenings** patients experienced constantly.

Lilian had been ingenious and resilient in the eleven or twelve years since her illness started. She had brought inner resources of every kind to her own aid: visual, musical, emotional, intellectual. Her family, her friends, her husband and daughter, above all, but also her students and colleagues, helpful people in the supermarket or on the street—everyone had helped her cope. Her adaptations to the agnosia were extraordinary—a lesson in what could be done to hold together a life in the face of ever-advancing perceptual and cognitive challenge. But it was in her art, her music, that Lilian not only coped with disease but transcended it. This was clear when she played the piano, an art that both demands and provides a sort of superintegration, a total integration of sense and muscle, of body and mind, of memory and fantasy, of intellect and emotion, of one's whole self, of being alive. Her musical powers, mercifully, remained untouched by her disease.

Her piano playing always added a transcendent note to my visits, and it recalled her, no less crucially, to her identity as an artist. It showed the joy she could still get and give, whatever other problems were now closing in on her.

When I revisited Lilian and Claude in 2002, I found the apartment full of balloons. "It was my birthday, three days ago," Lilian explained. She did not look well and seemed somewhat frail, although her voice and her warmth were entirely unchanged. She said that her visual powers had deteriorated further, and this was all too evident as she groped for a chair to sit down on, walked in the wrong direction, and got lost inside her own apartment. Her behavior now looked much more "blind," reflecting not only her increasing inability to decipher what faced her but a complete lack of visual orientation.

She was still able to write letters, but reading, even the painfully slow letter-by-letter reading that she could do a few years before, had become impossible. She adored being read to—Claude would read to her from newspapers and books—and I promised to send her some audiotapes. She could still go out a little, walking around the block on her husband's arm. The two of them were closer than ever, with her increasing disability.

Despite all this, Lilian felt that her ear was as good as it always was, and she had been able to continue a little teaching, with students from the music college coming to her apartment. Apart from this, though, she no longer played the piano much.

And yet, when I mentioned the Haydn quartet she had played for me before, her face lit up. "I was absolutely enthralled by that piece," she said. "I'd never heard it before. It's very rarely played." And she described for me again how, unable to get it out of her head, she had arranged it, mentally, for the piano, overnight. I asked her to play it for me again. Lilian demurred, and then, persuaded, started for the piano, but went in the wrong direction. Claude corrected her gently. At the piano, she first blundered, hitting wrong notes, and seemed anxious and confused. "Where am I?" she cried, and my heart sank. But then she found her place and began to play beautifully, the sound soaring up, melting, twisting into itself. Claude was amazed and moved by this. "She hasn't played at all for two or three weeks," he whispered to me. As she played, Lilian stared upward, singing the melody softly to herself. She played with consummate artistry, with all the power and feeling she had shown before, as Haydn's music swelled into a furious turbulence, a musical altercation. Then, as the quartet drew to its final, resolving chords, she said, simply, "All is forgiven."

Recalled to Life

PATRICIA H. was a brilliant and energetic woman who represented artists, ran an art gallery on Long Island, and was a talented amateur painter herself. She had raised her three children and, nearing sixty, continued to lead an active and even, as her daughters put it, "glamorous" life, with scouting expeditions to the Village and frequent soirées at home—she loved to cook, and there would often be twenty people for dinner. Her husband, too, was a man of many parts—a radio broadcaster, a fine pianist who sometimes performed at nightclubs, and politically active. Both were intensely sociable.

In 1989, Pat's husband died suddenly of a heart attack. Pat herself had had open-heart surgery for a damaged valve the year before, and had been put on anticoagulants. She had taken this in stride—but with her husband's death, as one of her daughters said, "She seemed stunned, became very depressed, lost weight, fell in the subway, had accidents with the car, and would show up, as if lost, on our doorstep in Manhattan." Pat had always been somewhat volatile in mood ("She would be depressed for a few days and take to her bed, then leap up in an opposite frame of mind and rush into the city, with a thousand engagements of one sort and another"), but now a fixed melancholy descended on her.

When, in January of 1991, she did not answer her phone for two days, her daughters became alarmed and called a neighbor, who, with the police, broke into Pat's house to find her lying in bed unconscious. She had been in a coma for at least twenty hours, the daughters were told, and had suffered a massive cerebral hemorrhage. There was a huge clot of blood in the left half of her brain, her dominant hemisphere, and it was thought that she would not survive.

After a week in the hospital with no improvement, Pat underwent surgery as a last-ditch measure. The results of this, her daughters were told, could not be predicted.

Indeed, it seemed at first, after the clot was removed, that the situation was dire. Pat would "stare . . . without seeming to see," according to one of her daughters. "Sometimes her eyes would follow me, or seem to. We didn't know what was going on, whether she was there." Neurologists sometimes speak of "chronic vegetative states," zombie-like conditions in which certain primitive reflexes are preserved but no coherent consciousness or self. Such states can be cruelly tantalizing, for there is often the feeling that the person is about to come to—but the states may last for months or even indefinitely. In Pat's case, though, it lasted for two weeks and then one day, as her daughter Lari recalled, "I had a Diet Coke in my hand—she wanted it. I saw her eye it. I asked, 'Do you want a sip?' She nodded. Everything changed at that moment."

Pat was conscious now, recognized her daughters, was aware of her condition and her surroundings. She had her appetites, her desires, her personality, but she was paralyzed on the right side, and, more gravely, she could no longer express her thoughts and feelings in words; she could only eye and mime, point or gesture. Her understanding of speech, too, was much impaired. She was, in short, aphasic.

"Aphasia" means, etymologically, a loss of speech, yet it is not speech as such which is lost but language itself—its expression or its comprehension, in whole or in part. (Thus, congenitally deaf people who use sign language may acquire aphasia following a brain injury or stroke and be unable to sign or understand sign language—a sign aphasia in every way analogous to the aphasia of speaking people.)

There are many different forms of aphasia, depending on which parts of the brain are involved, and a broad distinction is usually made between expressive aphasias and receptive aphasias—if both are present, this is said to be a "global" aphasia.

Aphasia is not uncommon; it has been estimated that one person in three hundred may have a lasting aphasia from brain damage, whether as the consequence of a stroke, a head injury, a tumor, or a degenerative brain disease. Many people, however, have a complete or partial recovery from aphasia. (There are also transient forms of aphasia, lasting only a few minutes, which may occur during a migraine or a seizure.)

In its mildest forms, expressive aphasia is characterized by a difficulty finding words or a tendency to use the wrong words, without compromise of the overall structure of sentences. Nouns, including proper names, tend to be especially affected. In more severe forms of expressive aphasia, a person is unable to generate full, grammatically complete sentences and is reduced to brief, impoverished, "telegraphic" utterances; if the aphasia is very severe, the person is all but mute, though capable of occasional ejaculations (such as "Damn!" or "Fine!"). Sometimes a patient may perseverate on a single word or phrase which is uttered in every circumstance, to their evident frustration. I had one patient who, after her stroke, could say nothing but "Thank you, Mama" and another, an Italian woman, who could utter only "Tutta la verità, tutta la verità."

Hughlings Jackson, a pioneer explorer of aphasia in the 1860s and '70s, considered that such patients lacked "propositional" speech, and that they had lost internal speech as well, so they could not speak or "propositionize," even to themselves. He felt therefore that the power of abstract thought was lost in aphasia, and in this sense, he compared aphasics to dogs.

In his excellent book **Injured Brains of Medical Minds,** Narinder Kapur cites many autobiographical accounts of aphasia. One of these is from Scott Moss, a psychologist who had a stroke at the age of forty-three, became aphasic, and later described his experiences, which were very much in accord with Hughlings Jackson's notions about the loss of inner speech and concepts:

> When I awoke the next morning in the hospital, I was totally (globally) aphasic. I could understand vaguely what others said to me if it was spoken slowly and represented a very concrete form of action. . . . I had lost completely the ability to talk, to read and to write. I even lost for the first two months the ability to use words internally, that is, in my thinking. . . . I had also lost the ability to dream. So, for a matter of eight to nine weeks, I lived in a total vacuum of self-produced concepts. . . . I could deal only with the immediate present. . . . The part of myself that was missing was [the] intellectual aspect—the sine qua non of my personality—those essential elements most important to being a unique individual. . . . For a long period of time I looked upon myself as only half a man.

Moss, who had both expressive and receptive aphasia, also lost the ability to read. For someone who has only an expressive aphasia, it may still be possible to read and to write (provided the writing hand is not paralyzed by the stroke).[1]

---

1. Macdonald Critchley described how Dr. Samuel Johnson lost all ability to speak when he suffered a stroke at the age of

Another account was that of Jacques Lordat, an eminent early-nineteenth-century French physiologist who provided an extraordinary description of his own aphasia after a stroke, sixty-odd years before Hughlings Jackson's studies. His experiences were quite different from Moss's:

> Within twenty-four hours all but a few words eluded my grasp. Those that did remain proved to be nearly useless, for I could no longer recall the way in which they had to be coordinated for the communication of ideas. . . . I was no longer able to grasp the

---

seventy-three. "In the middle of the night," Critchley wrote, "he awoke and immediately realized that he had sustained a stroke." To satisfy himself that he was not losing his sanity, Johnson composed a Latin prayer in his mind, but found he could not say it aloud. The next morning, June 17, 1783, he gave his servant a note he had been able to write for his next-door neighbor:

> Dear Sir, It hath pleased almighty God this morning to deprive me of the powers of speech; and, as I do not know that it may be his farther good pleasure to deprive me soon of my senses, I request you will, on the receipt of this note, come to me, and act for me, as the exigencies of my case may require.

Johnson continued writing letters, with his accustomed richness and magniloquence, over the next few weeks, while he slowly recovered the ability to speak. In some of the letters, though, he made uncharacteristic mistakes, sometimes omitting a word or writing the wrong word; he then corrected his mistakes upon rereading.

> ideas of others, for the very amnesia that prevented me from speaking made me incapable of understanding the sounds I heard quickly enough to grasp their meaning. . . . Inwardly, I felt the same as ever. This mental isolation which I mention, my sadness, my impediment and the appearance of stupidity which it gave rise to, led many to believe that my intellectual faculties were weakened. . . . I used to discuss within myself my life work and the studies I loved. Thinking caused me no difficulty whatever. . . . My memory for facts, principles, dogmas, abstract ideas, was the same as when I enjoyed good health. . . . I had to realize that the inner workings of the mind could dispense with words.

Thus in some patients, even if they are totally unable to speak or understand speech, there may be perfect preservation of intellectual powers—the power to think logically and systematically, to plan, to recollect, to anticipate, to conjecture.[2]

---

2. This was very much the case with Sir John Hale, the distinguished historian, who had a stroke that left him with expressive aphasia. His wife, Sheila Hale, in her book **The Man Who Lost His Language**, provides a vivid and moving account of her husband's aphasia, so devastating at first, and how he was able, partly through the power of expert and continuing therapy, to recover, even years later, much of what had seemed irreparably

Nevertheless, a feeling remains in the popular mind—and all too often in the medical mind, too—that aphasia is a sort of ultimate disaster which, in effect, ends a person's inner life as well as their outer life. Something along these lines was said to Pat's daughters, Dana and Lari. A little improvement, they were told, might occur, but Pat would need to be put away for the rest of her life; there would be no parties, no conversation, no art galleries anymore—all that had constituted the very essence of Pat's life would be gone, and she would lead the narrow life of a patient, an inmate, in an institution.

Scarcely able to initiate conversation or contact with others, patients with aphasia face special dangers in chronic hospitals or nursing homes. They may have therapy of every sort, but a vital social dimension of their lives is missing, and they frequently feel intensely isolated and cut off. Yet there are many activities—card games, shopping trips, movies or theater, dancing or sports—that do not require language, and these can be used to draw or inveigle aphasic patients into a world of familiar activities

lost. And she brings out how even medical professionals may dismiss aphasic patients as "incurable" or treat them as idiots, despite their manifest intelligence.

and human contact. The dull term "social rehabilitation" is sometimes used here, but really the patient (as Dickens might put it) is being "recalled to life."

Pat's daughters were determined to do everything they could to bring their mother back into the world, to the fullest possible life her limitations allowed. "We hired a nurse who retaught my mother how to feed herself, how to **be**," Lari said. "Mother would get angry, sometimes strike her, but she, the nurse, would never give up. Dana and I never left her side. We would take her out, wheel her to my apartment. . . . We would take her out to restaurants, or bring food in, have her hair done, her nails manicured. . . . We never stopped."

Pat was moved from the acute care hospital where she had had surgery to a rehabilitation facility. After six months, she was finally moved to Beth Abraham Hospital, in the Bronx, where I first met her.

When Beth Abraham Hospital was opened, in 1919, it was called the Beth Abraham Home for Incurables, a discouraging name that was changed only in the 1960s. Originally accommodating some of the first victims of the encephalitis lethargica epidemic (some of whom were still living there more than forty years later, when I arrived), Beth Abraham expanded over the years to become a five-hundred-bed hospital with active rehabilitation programs aimed at helping patients with all sorts of chronic conditions: parkinsonism, dementias, speech problems, multiple sclerosis, strokes (and, increasingly, spinal or brain damage from bullet wounds or car accidents).

Visitors to hospitals for the chronically ill are often horrified at the sight of hundreds of "incurable" patients, many of them paralyzed, blind, or speechless. One's first thought is often: Is life worth living in circumstances like these? What sort of a life can these people have? One wonders, nervously, how one would react to the prospect of being disabled and entering such a home oneself.

Then one may start to see the other side. Even if no cure, or only limited improvement, is possible for most of these patients, many of them can nonetheless be helped to reconstruct their lives, to develop other ways of doing things, capitalizing on their strengths, finding compensations and accommodations of every sort. (This, of course, depends upon the degree and type of neurological damage, and upon the inner and outer resources of the individual patient.)

If the first sight of a chronic hospital can be hard for visitors, it can be terrifying for a new inmate; many react with horror mixed with sadness, bitterness, or rage. (Sometimes this even results in a full-blown "admission psychosis.") When I first met Pat, shortly after her admission to Beth Abraham in October of 1991, I found her angry, tormented, and frustrated. She did not yet know the staff or the layout of the place, and she felt that a rigid, institutional order was being imposed on her. She could communicate through gestures—these were passionate, if not always understandable—but she still had no coherent speech (though occasionally, the staff said, she would exclaim "Hell!" or "Go away!" when she was angry). While she seemed to understand a good deal of what people said to her, it became clear, on examination, that she was responding not so much to words as to the tone of voice, facial expression, and gestures.

When I tested her in the clinic, Pat could not respond to "Touch your nose," either in speech or in writing. She could count ("one, two, three, four, five . . .") as a sequence, but could not say individual numbers or count backward. The right side of her body remained completely paralyzed. Her neurological situation, I noted in my report, was "a bad one. I fear there may not be too much recovery of language functions, but intensive speech therapy, as well as physical therapy and occupational therapy, must certainly be tried."

Pat yearned to speak, but was continually frustrated when, after huge efforts to get a word out, it would be the wrong word, or unintelligible. She would try to correct it, but often would become more unintelligible with every attempt to make herself understood. It started to dawn on her, I think, that her power of speech might never come back, and increasingly she retreated into silence. This inability to communicate was, for her, as for many patients with aphasia, far worse than the paralysis of half her body. I would sometimes see her, in this first year after her stroke, sitting alone in the corridor or in the patients' dayroom, bereft of speech, surrounded by a sort of penumbra of silence, with a stricken and desolate look on her face.

But a year later, I found Pat much improved. She had developed a knack for understanding other people by their gestures and expressions as much as their words. She could indicate her own thoughts and feelings not by speech but by eloquent gesture and mime. She indicated, for example, fluttering a couple of tickets, that she would go to the movies if, and only if, a friend of hers could go, too. Pat had become less angry, more sociable, and very aware of all that was going on around her.

This represented an enormous social improvement—an improvement in her ability to communicate—but I was not sure how much it rested on actual neurological improvement. Friends and relatives of aphasic patients often think that there is more neurological recovery than there actually is, because many such patients can develop a remarkable compensatory heightening of other, nonlinguistic powers and skills, especially the ability to read others' intentions and meanings from their facial expressions, vocal inflections, and tone of voice, as well as all the gestures, postures, and minute movements that normally accompany speech.

Such compensation may give surprising powers to the aphasic—in particular, an enhanced ability to see through histrionic artifice, equivocation, or lying. I described this in 1985,[3] when I observed a group of aphasic patients watching a presidential speech on television, and in 2000 Nancy Etcoff and her colleagues at Massachusetts General Hospital published a study in **Nature** which showed that people with aphasia were in fact "significantly better at detecting lies

3. "The President's Speech," a chapter in **The Man Who Mistook His Wife for a Hat.**

about emotion than people with no language impairment." Such skills, they observed, apparently took time to develop, for they were not evident in a patient who had been aphasic for only a few months. This seemed to be the case with Pat, who had initially been far from expert in picking up others' emotions and intentions but over the years had become preternaturally skillful at it. If aphasic people come to excel in understanding nonverbal communication, they can also become expert in conveying their own thoughts in the same way—and Pat was now starting to move towards a conscious and voluntary (and often inventive) representation of her thoughts and intentions by mime.

Yet while gesture and mime, lacking the grammar and syntax of real language, are usually spared in aphasia, they are not enough; they have only a limited ability to convey complex meanings and propositions (unlike a true sign language, such as deaf people use). These limitations often infuriated Pat, but a crucial change came when her speech pathologist, Jeannette Wilkens, discovered that though Pat could not read a sentence, she could recognize individual words (and that, indeed, her vocabulary was quite extensive). Jeannette had found this with other aphasic patients as they started to recover, and she had devised a sort of lexicon for them, a book of words arranged in categories of objects, people, and events, as well as moods and emotions.

Such a lexicon often worked, Jeannette found, when patients were closeted in one-on-one sessions with her, but many aphasic patients had difficulty reaching out to others—perhaps they were too shy, too depressed, or too disabled from other medical conditions to initiate contact with other people.[4] None of this was the case with Pat, who had been outgoing and social all her life. She always carried the book on her lap or at the side of her wheelchair, so she could leaf through it rapidly with her left hand and find the words she needed. She would boldly approach someone, open her book to the right page, thrust it at them, and point to the subject she wanted to talk about.

Pat's life expanded in all sorts of ways with her "bible," as her daughters called it. Soon she was able to guide a conversation in any direction she wanted, a conversation that was on

4. Some of Wilkens's extraordinary therapeutic powers may have gone with the fact that she herself was quadriplegic (having broken her neck in a car accident at the age of eighteen) but nevertheless led an extremely full life and was deeply interested in other people. Seeing the fortitude and resilience of a therapist in some ways even more disabled than themselves inspired Wilkens's patients to work harder for her, and for themselves.

her part accomplished solely by gesture and mime—and this had to be done primarily with her left arm, for her right side was still completely paralyzed. Nevertheless, the combination of gesture and mime with the words in her book allowed her a remarkably full and exact expression of her needs and thoughts.

Inside the hospital, she became a central social figure, despite being unable to communicate in the usual way. Her room became a chat room, with other patients often dropping by. Pat would talk to her daughters on the phone, they said, "a hundred times a day," though the conversations were all passive on her part, awaiting simple questions to which she could answer "yes" (she communicated "yes" by kisses), "no," or "fine," or by noises of approbation, amusement, or disapproval.

By 1996, five years after her stroke, Pat's receptive aphasia had lessened; she was able to understand a little speech, though still unable to express herself in speech. She had certain fixed phrases, like "You're welcome!" or "Fine!," but could not name familiar objects or utter a sentence. She started to paint once again, using her left hand, and she was a terror at dominoes—her nonverbal representational systems were unimpaired. (It has long been understood that aphasia need not affect musical ability, visual imagery, or mechanical aptitude, and Nicolai Klessinger and his colleagues at the University of Sheffield have shown that numerical reasoning and mathematical syntax can be entirely intact even in patients who are unable to understand or produce grammatical language.)

It is often said that following a stroke or a brain injury, no further recovery is possible after twelve to eighteen months. While this may sometimes be so, I have seen this generalization proved false in many individual patients. And in the past few decades neuroscience has confirmed that the brain has more powers of repair and regeneration than was once believed. There is far more "plasticity," too, a greater capacity for undamaged brain areas to take over some of the functions of damaged ones, provided the damage is not too extensive. And at a personal level, there are powers of accommodation: finding new ways or other ways of doing things when the original way is no longer available. Even five years after her stroke, I noted that Pat was still showing a continuing, though very limited, improvement in her receptive powers, her ability to understand language.

Nonetheless, despite her ability to ejaculate a few words, and her ability to understand single words whether spoken or written, Pat was still, basically, bereft of organized language and seemed unable to "propositionize" either internally or to others. Wittgenstein, the philosopher, distinguished two methods of communication and representation: "saying" and "showing." Saying, in the sense of propositionizing, is assertive and requires a tight coupling of logical and syntactic structure with what it asserts. Showing is not assertive; it presents information directly, in a nonsymbolic way, but, as Wittgenstein was forced to concede, it has no underlying grammar or syntactic structure. (A few years after Wittgenstein's **Tractatus** was published, his friend Piero Sraffa made a gesture, snapped his fingers, and said, "What is the logical structure of that?" Wittgenstein could not answer.)

As Noam Chomsky revolutionized the study of language, Stephen Kosslyn has revolutionized the study of imagery, and where Wittgenstein writes of "saying" and "showing," Kosslyn speaks of "descriptive" and "depictive" modes of representation. These modes are both available to the normal brain, and they are complementary, so that one may sometimes use one mode or the other, and often both together. Pat had largely lost her powers of propositionizing, of asserting, of describing, and showed little likelihood of regaining these. But her powers of depiction, spared by the stroke, were remarkably heightened in reaction to her loss of language. Her power to read others' gestures and expressions and her virtuosity in expressing herself through gesture and mime constituted the two sides—receptive and expressive—of her depictive power.

Pat was the youngest of seven siblings; her extended family had always played a central part in her life, and this extended further still when Lari's daughter Alexa, Pat's first grandchild, was born in 1993. Alexa, said Lari, "was born into Beth Abraham." She would visit her grandmother frequently, and Pat always had a special toy or treat for her ("I don't know how she got these things," Lari marveled). Pat would often ask Alexa to take crackers to a friend down the hall who could not walk. Alexa and her two younger siblings, Dean and Eve, were all fascinated by Pat and liked to call her often on the phone when they were unable to visit her. Lari felt that they had a very active, very "normal" relationship with their grandmother, a relationship they all treasured.

One of the pages in Pat's book contained a list of emotional states (she had picked these out from a word list prepared by Jeannette, the speech pathologist). When I asked her, in 1998, what her predominant mood was, she pointed to "happy." There were other adjectives on the mood page, such as "furious," "scared," "tired," "sick," "lonely," "sad," and "bored"—all of which she had indicated, on occasion, in previous years.

In 1999, when I asked her the date, she pointed to "Wednesday, July 28"—a little miffed, perhaps, that I had insulted her with such a simple question. She indicated, using her "bible," that she had been to half a dozen musicals and a couple of art galleries in the past few months, and that, now that it was summer, she would visit Lari on Long Island on the weekends and, among other things, swim. "Swim?" I asked, incredulous. Yes, Pat indicated; even with her right side paralyzed, she could still do the sidestroke. She had been a great long-distance swimmer, she indicated, in her youth. She told me how excited she was that Lari would be adopting a new baby in a few months. I was especially struck, on this visit, eight years after her stroke, by the fullness and richness of Pat's daily experiences and her voracious love of life in the face of what one might judge to be devastating brain damage.

In 2000, Pat showed me photos of her grandchildren. She had visited them all the previous day, for the Fourth of July, and they had watched the tall ships and the fireworks on television. She was eager to show me the newspaper, with a picture of the Williams sisters playing tennis. Tennis, she indicated, had been one of her favorite sports, along with skiing, riding, and swimming. She was at pains to show me that her fingernails were manicured and painted, and she was dressed in a sun hat and sunglasses, on her way to sun herself on the hospital patio.

By 2002, Pat had become able to use a few spoken words. This was achieved by the use of familiar songs like "Happy Birthday" or "A Bicycle Built for Two," which she would sing along with Connie Tomaino, Beth Abraham's music therapist. Pat was able to get the feeling of the music and some of the words. For a few minutes afterward, this would "release" her voice and give her the ability to say some of the words, in a singsong fashion. She started carrying a tape recorder with a cassette of familiar songs, so she could get her language powers working. She demonstrated this with "Oh, What a Beautiful Morning," followed by a melodious "Good morning, Dr. Sacks," with a heavy, rhythmic emphasis on "morning."

Music therapy is invaluable for some patients with expressive aphasia; finding that they can sing the words to a song, they are reassured that language is not wholly lost, that they still have access to words somewhere inside them. The question is then whether the language capacities embedded in song can be removed from their musical context and used for communication. This is sometimes possible to a limited extent, by reembedding words in a sort of improvised singsong.[5] But Pat's heart was not in this—she felt that her real virtuosity lay in her mimetic powers, her appreciation and use of gesture. She had achieved a skill and intuitiveness here amounting almost to genius.

Mimesis, the deliberate and conscious representation of scenes, thoughts, feelings, intentions, and so on, by mime and action, seems to be a specifically human achievement, like language (and perhaps music). Apes, which are able to "ape," or imitate, have little power to create conscious and deliberate mimetic representations. (In **Origins of the Modern Mind,** the psychologist Merlin Donald suggests that a "mimetic culture" may have been a crucial intermediate stage in human evolution, between the "episodic" culture of apes and the "theoretic" culture of modern man.) Mimesis has a

---

5. I have written more extensively about music therapy for aphasia in a chapter of **Musicophilia.**

much larger and more robust cerebral representation than language, and this may explain why it is so often preserved in patients who have lost language. This preservation can allow remarkably rich communication, especially if it can be elaborated and heightened and combined, as in Pat's case, with a lexicon.

Pat had always had a passion to communicate ("This was a woman who talked twenty-four hours a day," Dana said), and it was the frustration of this loquacity that led to despair and fury when she first arrived at the hospital, and to her intense motivation and success in communicating once Jeannette got her going.

Pat's daughters were sometimes amazed at her resilience. "Why isn't she depressed," Dana said, "given her earlier history of depression? How could she live like this, I thought at first. . . . I thought she would take a knife to herself." Every so often, Dana related, her mother would make a gesture that seemed to say, "My God, what happened? What is this? Why am I in this room?" as if the raw horror of her stroke had hit her once again. But Pat was aware that she had, in a sense, been very lucky, even though half of her body remained paralyzed. She was lucky that her brain damage, though extensive, did not undermine her force of mind or personality; lucky that her daughters fought so hard from the beginning to keep her engaged and active and were able to afford extra aides and therapists; lucky, too, that she encountered a speech pathologist who observed her sensitively and minutely, one who was so personally inspiring and could provide her with a crucial tool, her "bible," which worked so well.

Pat continued to remain active and engaged with the world. She was, as Dana said, the "darling" of the family, and of the floor at the hospital, too. She had not lost the power to captivate people ("She has even captivated you, Dr. Sacks," Dana observed), and she could do a little painting with her left hand. She was grateful to be alive and to be able to do as much as she could, and this, Dana thought, was why her mood and morale were so good.

Lari expressed herself in similar terms. "It's as if the negativity has been wiped away," she told me. "She is much more consistent, appreciative of her life and gifts . . . of other people, too. She is conscious of being privileged, but this makes her kinder, more thoughtful to other patients who may be physically less disabled than she is but much less 'adapted' or 'lucky' or 'happy.' She is the opposite of a victim," Lari concluded. "She actually feels that she has been blessed."

. . .

One cool Saturday afternoon in November, I joined Pat and Dana for one of Pat's favorite activities: shopping on Allerton Avenue, near the hospital. When we arrived in Pat's room—it was overflowing with plants, paintings, photos and posters, theater programs—Pat was awaiting us, already wearing a favorite coat.

As we went up Allerton Avenue, bustling on a weekend afternoon, I saw that half the shopkeepers knew Pat; they shouted "Hi, Pat!" as she bowled past in her wheelchair. She waved at the young woman in the health food store where she buys her carrot juice, and received a "Hi, Pat!" back. She waved to a Korean woman at the dry cleaner's, blew a kiss, and had a kiss blown back. The woman's sister, Pat was able to indicate to me, used to work in the fruit store. We entered a shoe shop, where Pat's desires were very clear: she wanted a boot with fur inside, for the upcoming winter. "Zip or Velcro?" Dana inquired. Pat indicated no preference, but wheeled herself in front of the boot display and then, with great decisiveness, pointed to the boots she wanted. Dana said, "But they have laces!" Pat smiled and shrugged, meaning, "So what! Someone else will tie them." She is not without vanity—the boots had to be elegant as well as warm. ("Velcro, indeed!" her expression said.) "What size? A nine?" Dana asked. No, Pat gestured, bisecting her finger; an eight and a half.

We stopped by the supermarket, where she always picks up a few things for herself and for others at the hospital. Pat knew every aisle and quickly picked two ripe mangoes for herself, a large bunch of bananas (most, she gestured, she would give away), some small doughnuts, and, at the checkout, three bags of candy. (She indicated that these were for the children of an orderly on her floor.)

As we moved on, laden with our purchases, Dana asked me where I had been earlier in the day. I said I had been to a meeting of the Fern Society at the New York Botanical Garden, adding, "I'm a plant person." Pat, overhearing, made a wide gesture and pointed to herself, meaning, "You and I. We are both plant people."

"Nothing has changed since her stroke," Dana said. "She has all her old loves and passions. . . . The only thing is," she added, smiling, "she has become a pain in the neck!" Pat laughed, agreeing with this.

We stopped at a coffee shop. Pat clearly had no difficulty with the menu, indicating that she wanted not home fries but French fries, with whole wheat toast. After the meal, Pat carefully applied lipstick. ("How vain!" Dana exclaimed, with admiration.) Dana wondered whether she could take her mother on a cruise. I mentioned the giant cruise ships I had seen go in and out of Curaçao, and Pat, intrigued, inquired with her book whether they set out from New York. I tried to draw a ship in my notebook; Pat laughed and, left-handed, did a much better one.

A Man of Letters

In January of 2002, I received a letter from Howard Engel, the Canadian writer known for his Benny Cooperman series of detective novels, describing a strange problem. One morning a few months before, he wrote, he had got up feeling fine. He dressed and made breakfast and then went to the front porch to get his newspaper. But the paper on his doorstep seemed to have undergone an uncanny transformation:

> The July 31, 2001, **Globe and Mail** looked the way it always did in its make-up, pictures, assorted headlines and smaller captions. The only difference was that I could no longer read what they said. The letters, I could tell, were the familiar twenty-six I had grown up with. Only now, when I brought them into focus, they looked like Cyrillic one moment and Korean the next. Was this a Serbo-Croatian version of the **Globe**, made

> for export? . . . Was I the victim of a practical joke? I have friends who are capable of such things. . . . I wondered what I might do to them that would improve on this piece of foolery. Then, I considered the alternative possibility. I checked the **Globe**'s inside pages to see if they looked as strange as the front page. I checked the want ads and the comics. I couldn't read them either. . . .
>
> Panic should have hit me like the proverbial ton of bricks. But instead I was suffused with a reasonable, business-as-usual calm. "Since this isn't somebody's idea of a joke, then, it follows, I have suffered a stroke."

Along with this realization came a memory of a case history he had read a few years earlier, my own "Case of the Colorblind Painter."[1] He remembered in particular how my patient, Mr. I., following a head injury, found himself unable to read the police accident report—he saw print of different sizes and types but could make nothing of it, and said it looked "like Greek or Hebrew." He remembered, too, that Mr. I.'s inability to read, his alexia, had lasted for five days and then cleared.

Howard kept testing himself, turning over the pages, to see if everything would suddenly

1. This was published as a chapter in **An Anthropologist on Mars.**

snap back to normal. Then he went into his library; maybe, he thought, "books would behave better than the newspaper." The room looked normal, and he noted that he could still read his clock, but his books—some in French and German, as well as English—were all unintelligible, all full of the same "Oriental"-looking script.

He woke his son, and together they took a cab to the hospital. Along the way, Howard thought he saw "familiar landmarks in unfamiliar places," and he could not read the names of streets as they passed, nor the words "Emergency Room" when they arrived at the hospital—though he at once recognized the picture of an ambulance over the door. He underwent a battery of tests, and these confirmed his own suspicion: he had indeed had a stroke; he was told that it affected a limited area of the visual parts of the brain, on the left side. During the intake interview at the hospital, he later recalled, he was somewhat confused: "I was unable to pinpoint my exact relationship to my son. . . . I forgot my name, my age, my address, and a dozen other things."

Howard spent the next week in the neurology ward at Toronto's Mount Sinai Hospital. During this time it became clear that he had other visual problems besides his inability to read: he had a large blind spot in the upper right quadrant of his visual field, and he had difficulties recognizing colors, faces, and everyday objects. These difficulties would come and go, he noted:

> Familiar objects like apples and oranges suddenly look[ed] strange, as unfamiliar as an exotic piece of Asian fruit. A rambutan. I would surprise myself with not knowing whether I was holding an orange or a grapefruit, a tomato or an apple. Usually, I could sort them out by sniffing or squeezing.

He often forgot things he once knew perfectly well, and became shy of conversation, he wrote, "lest I forget the name of the prime minister or who wrote **Hamlet.**"

Yet he was surprised to find, as a nurse reminded him, that he could still **write,** even though he could not read; the medical term, she said, was "alexia sine agraphia." Howard was incredulous—surely reading and writing went together; how could he lose one but not the other?[2] The nurse suggested that he sign

---

2. Lilian Kallir, too, had alexia sine agraphia, and continued to write letters to her friends around the world. But since her alexia for words developed slowly, over the course of years, she seemed to have insensibly accommodated to the fact that reading and writing could be dissociated in this way.

his name; he hesitated, but once he started, the writing seemed to flow all by itself, and he followed his signature with two or three sentences. The act of writing seemed quite normal to him, effortless and automatic, like walking or talking. The nurse had no difficulty reading what he had written, but he himself could not read a single word. To his eyes, it was the same indecipherable "Serbo-Croatian" he had seen in the newspaper.

We think of reading as a seamless and indivisible act, and as we read we attend to the meaning and perhaps the beauty of written language, unconscious of the many processes that make this possible. One has to encounter a condition such as Howard Engel's to realize that reading is, in fact, dependent on a whole hierarchy or cascade of processes, which can break down at any point.

In 1890, the German neurologist Heinrich Lissauer used the term "psychic blindness" to describe how some patients, after a stroke, became unable to recognize familiar objects visually.[3] People with this condition, visual agnosia,

3. The current term, "visual agnosia," was introduced by Sigmund Freud the following year.

can have perfectly normal visual acuity, color perception, visual fields, and so on—yet be totally unable to recognize or identify what they are seeing.

Alexia is a specific form of visual agnosia, an inability to recognize written language. Since the French neurologist Paul Broca in 1861 had identified a center for the "motor images" of words, as he called it, and his German counterpart Carl Wernicke, a few years later, identified one for the "auditory images" of words, it seemed logical to nineteenth-century neurologists to suppose that there might also be an area in the brain dedicated to the **visual** images of words—an area that, if damaged, would produce an inability to read, a "word blindness."[4]

---

4. Congenital "word blindness" (which we now call dyslexia) was recognized by neurologists in the 1880s, around the same time that Charcot, Déjerine, and others were describing acquired alexia. Children with severe difficulties in reading (sometimes in writing, reading music, or calculating, too) were often seen as retarded, despite clear evidence to the contrary. W. Pringle Morgan, writing in the **British Medical Journal** in 1896, detailed a careful study of an intelligent and articulate fourteen-year-old boy who had severe difficulties in reading and spelling:

> In writing his own name he made a mistake, putting "Precy" for "Percy," and he did not notice the mistake until his attention was called to it more than once. . . . Words written or printed seem to convey no impression

In 1887, a French neurologist, Joseph-Jules Déjerine, was asked by an ophthalmologist colleague to see a highly intelligent, cultivated man who had suddenly lost the ability to read. Edmund Landolt, the ophthalmologist, wrote a short but vividly evocative portrait of the patient, and Déjerine, in his own paper on the subject, included a long excerpt from this.

They described how in October of that year, Oscar C., a retired businessman, found himself suddenly unable to read. (He had had some brief attacks of numbness in his right leg on previous days, but had paid little attention to them.) Though reading was impossible, Monsieur C. had no difficulty recognizing people

---

> to his mind, and it is only after laboriously spelling them that he is able, by the sounds of the letters, to discover their import. . . . He can only recognize such simple ones as "and," "the," "of," etc. Other words he never seems to remember, no matter how frequently he may have met them. . . . The schoolmaster who has taught him for some years says that he would be the smartest lad in the school if the instruction were entirely oral.

It is now recognized that as much as five to ten percent of the population has dyslexia and that, whether by way of "compensation" or simply because of their different neurological makeup, many dyslexic people have exceptional talents in other areas. These and many other aspects of dyslexia are discussed in depth by Maryanne Wolf in **Proust and the Squid: The Story and Science of the Reading Brain,** and by Thomas G. West in **In the Mind's Eye.**

and objects around him. Nevertheless, thinking that his eyes must be at fault, he consulted Landolt, who wrote:

> Asked to read an eye chart, C is unable to name any letter. However, he claims to see them perfectly. He instinctively sketches the form of the letters with his hand, but he is nevertheless unable to say any of their names. When asked to write on a paper what he sees, he is able, with great difficulty, to recopy the letters, line by line, as if he were making a technical drawing, carefully examining each stroke in order to reassure himself that his drawing is exact. In spite of these efforts, he remains incapable of naming the letters. He compares the **A** to an easel, the **Z** to a serpent, and the **P** to a buckle. His incapacity to express himself frightens him. He thinks that he has "gone mad," since he is well aware that the signs he cannot name are letters.[5]

Like Howard Engel, Monsieur C. was unable to read even the headlines of his morning paper, although he nonetheless recognized it, by its format, as his usual newspaper, **Le**

---

5. I am quoting here and elsewhere from the translation provided by Israel Rosenfield in his excellent book **The Invention of Memory.**

Matin. And, like Howard, he could write perfectly well:

> While reading is impossible, the patient . . . can write fluently and without any mistakes whatever material is dictated to him. But should he be interrupted in the middle of a phrase that he is writing . . . he becomes muddled and cannot start up again. Also, if he makes a mistake he can't find it. . . . He can never reread what he has written. Even isolated letters do not make sense to him. He can only recognize them . . . by tracing the outlines of the letter with his hand. Therefore it is the sense of the muscular movement that gives rise to the letter name. . . .
>
> He is able to do simple addition, since he recognizes, with relative ease, numbers. However, he is very slow. He reads the numbers poorly, since he cannot recognize the value of several numbers at once. When shown the number 112, he says, "It is a 1, a 1, and a 2," and only when he writes the number can he say "one hundred and twelve."[6]

---

6. Israel Rosenfield also remarks that Oscar C.'s central problem was not just in recognizing letters but in perceiving their sequence, and that he had similar problems with numerical notation. Numbers, Rosenfield writes, "are always read the same way in every context. A **3** is **three** whether it appears in the phrase '3 apples' or 'a 3 percent discount.' But . . . the mean-

There were some additional visual problems—objects appeared dimmer and a little blurred on the right side and completely devoid of color. These problems, along with the specificity of Oscar C.'s alexia, indicated to Landolt that the underlying problem was not in the eyes but in the brain; this led him to refer his patient to Déjerine.

Déjerine was fascinated by Monsieur C.'s condition and arranged to see him twice weekly at his clinic in Paris. In a monumental 1892 paper, Déjerine summarized his neurological findings succinctly and then, in a much more leisurely style, provided a general picture of his patient's life:

> C spends his days taking long walks with his wife. He has no difficulty walking and every

---

ing of a number in a multidigit numeral depends on where it is placed." It is similar with musical notes, whose meaning depends on context and placement.

Words, Rosenfield continues, are similar:

> Changing a single letter in a word can alter both its pronunciation and its meaning. Its significance depends on what precedes and what follows. . . . It is the failure to capture this overall organization—in which identical stimuli, letters, are constantly changed in significance—that is characteristic of patients with verbal blindness. They cannot organize the stimuli in a way that makes sense of the symbols.

> day he does his errands on foot from the Boulevard Montmartre to the Arc de Triomphe and back. He is aware of what is happening around him, stops in front of stores, looks at paintings in gallery windows, etc. Only posters and signs in shops remain meaningless collections of letters for him. He often becomes exasperated by this, and though he has been so afflicted for four years, he has never accepted the idea that he cannot read, while remaining able to write. . . . In spite of patient exercises and much effort, he has never relearned the sense of letters and written words, nor has he ever relearned how to read musical notes.

Despite this, Oscar C., an excellent singer, could still learn new music by ear, and he continued to practice music with his wife every afternoon. And he continued to enjoy and excel at playing cards: "He is a very good card player, calculates very well, prepares his blows well in advance and wins most of the time." (Déjerine did not comment on how Monsieur C. was able to "read" the cards, but it seems likely that he recognized the iconic images of hearts, diamonds, spades, clubs, jacks, queens, and kings—just as Howard Engel recognized the icon of an ambulance when he arrived at the emergency room. Number cards, of course, can also be recognized by their patterns.)

When Oscar C. died following a second stroke, Déjerine performed an autopsy and found two lesions in the brain: an older one, which had destroyed part of the left occipital lobe and which he presumed was responsible for Monsieur C.'s alexia, and a larger, recent lesion, which had probably caused his death.[7]

It is always difficult to make inferences from the appearance of the brain at autopsy; one may find damaged areas, but it is not always possible to see their manifold connections with other areas of the brain or to determine what controls what. Déjerine was well aware of this; nonetheless, he felt that by relating a specific neurological symptom—alexia—to damage in a particular area of the brain, he had, in principle, demonstrated what he called a "visual center for letters" in the brain.

Déjerine's discovery of this area essential for

---

7. In the few days that he lived after his second stroke, Oscar C. had aphasia as well. He would say one word in place of another or make garbled sounds and had to rely on mime and gesture to communicate. His wife noticed ("with dread") that he could no longer write. Israel Rosenfield, analyzing Déjerine's case in **The Invention of Memory**, suggests that one may have alexia without agraphia—this is relatively common—but not agraphia without alexia. "Agraphia," Rosenfield writes, "is always associated with an inability to read." And yet extremely rare cases of isolated agraphia have been reported, and the debate is not yet resolved.

reading would be confirmed over the next hundred years by scores of similar cases and autopsy reports of patients with alexia, irrespective of its cause.

By the 1980s, CT scanning and MRIs made it possible to visualize living brains with an immediacy and precision impossible in autopsy studies (where all sorts of secondary changes may blur the picture). Using this technology, Antonio and Hanna Damasio and, later, other researchers, were again able to confirm Déjerine's findings, and to correlate their alexic patients' symptoms with highly specific brain lesions.

With the development of functional brain imaging a few years later, it became possible to visualize the activity of the brain in real time, as subjects performed various tasks. A pioneer PET scan study in 1988 by Steven Petersen, Marcus Raichle, and their colleagues showed the different areas of the brain activated by reading words, listening to words, uttering words, and associating words. "For the first time in history," as Stanislas Dehaene writes in his book **Reading in the Brain**, "the areas responsible for language had been photographed in the living human brain."

Dehaene, a psychologist and neuroscientist, has specialized in studying the processes involved in visual perception, especially the recognition and representation of words, letters, and numbers. Using fMRI technology, which is much swifter and more sensitive than PET scanning, he and his colleagues have been able to focus even more closely on what he calls the visual word form area or, more informally, "the brain's letterbox."

Dehaene's studies (with Laurent Cohen and others) have shown how the visual word form area can be activated in a fraction of a second by a single written word, and how this initial, purely visual activation then spreads to other areas of the brain—especially the temporal lobes and the frontal lobes.

Reading, of course, does not end with the recognition of visual word forms—it would be more accurate to say that it begins with this. Written language is meant to convey not only the sound of words but their meaning, and the visual word form area has intimate connections to the auditory and speech areas of the brain as well as to the intellectual and executive areas, and to the areas subserving memory and emotion.[8] The visual word form area is a crucial

---

8. Kristen Pammer and her colleagues have also shown, using magnetoencephalography, that the visual word form area does not work in isolation; it is part of a widely distributed cerebral

node in a complex cerebral network of reciprocal connections—a network peculiar, it seems, to the human brain.

As a prolific writer and an omnivorous reader, accustomed to reading newspapers every morning and many books each week, Howard Engel wondered how he would manage life with his alexia, which showed no signs of clearing. In a world full of traffic signs, printed labels, and directions on everything from a prescription bottle to the television, ordinary life is a continuing, daily struggle for anyone with alexia. But for Howard, this was an even more desperate situation, for his whole life and identity (to say nothing of his livelihood) depended on his ability to read and write.

Being able to write without reading might be

---

network. Indeed, some areas in the frontal and temporal lobes are activated by words **before** the visual word form area. They stress that the spread of activation flows in both directions, to and from the visual word form area.

Nonetheless it is possible to separate the act of reading from meaning, as, for example, I do when I read a religious text in Hebrew. I have learned how the words sound, but have little idea of their meaning. Something similar happens with hyperlexic preschool children, usually autistic, who may be able to read an article in the **New York Times** fluently and correctly but without comprehension.

all right for a short letter or memorandum, a page or two. But for the most part, he thought, it "was like being told that the right leg had to be amputated but that I could keep the shoe and sock." How could he hope to go back to his previous work—to write an elaborate narrative of crime and detection, full of plots and counterplots, to do all the corrections and revisions and redrafting a writer must do—without being able to read? He would have to get others to read for him, or perhaps get one of the ingenious new software programs that would allow him to scan what he had written and hear it read back to him by a computer. Both of these would involve a radical shift from the visuality of reading, the look of words on a page, to an essentially auditory mode of perception—going, in effect, from reading to listening and, perhaps, from writing to speech. Would this be desirable—or even possible?

Precisely this question had forced itself on another writer who consulted me ten years earlier. Charles Scribner, Jr., was also a man of letters; he presided over the publishing house established by his great-grandfather in the 1840s. In his sixties, he developed a visual alexia—probably as a result of a degenerative process in the visual parts of the brain. It was a devastating problem for a man who had published the work of Hemingway and others, a man whose life was centered on reading and writing.

As a book publisher, Scribner slightly disapproved of audiobooks, which had recently been introduced to the general public. But he decided nonetheless to reconstruct his entire literary life in an auditory mode. To his surprise, this did not prove as difficult as he expected. He even began to enjoy listening to audiobooks:

> It never dawned on me that these spoken books would become a major part of my intellectual life and recreational reading. By now I must have "read" hundreds of books in this way. I was never a rapid reader as a boy, although my retention was high. Paradoxically, now that I was reading books on tape, my reading speed was better than ever and my retention just as good. I can fairly say that for me the discovery of this mode of reading was a kind of "open sesame" to my continued enjoyment of literature.[9]

Like Howard, Scribner preserved the power to write, but he was so deeply distressed by his

---

9. When we met, Scribner gave me a brief memoir he had just dictated, describing his alexia and how he had adapted to it; subsequently he published this as an afterword in his last book, **In the Web of Ideas**, from which I am quoting here.

inability to read what he had written that he decided to change to dictation, something he had never before tried. Luckily, this too was successful—dictation worked so well that it allowed him to complete more than eighty newspaper columns and two book-length memoirs about his life in publishing. "Perhaps," he wrote, "it's another instance of a handicap honing a skill." Apart from his close friends and family, no one seemed aware that he had accomplished all this by switching to an entirely new mode.

One might have expected Howard, too, to turn to an auditory mode of "reading" and writing, but his course was very different.

After his week at Mount Sinai Hospital, he was moved to a rehabilitation hospital, where he spent almost three months studying himself, what he could and could not do. When he was not trying to read a paper or a get-well card, he found, he could forget about his alexia:

> The sky looked blue, the sun shone on the hospital windows, the world hadn't suddenly become unfamiliar. My alexia existed only when I had my head buried in a book. Print brought it on and reminded me that, yes, there was a problem. Thus was born the temptation to simply avoid reading.

But this, he quickly realized, was unacceptable to him as a reader and a writer. Audiobooks might do for some, but not for him. He still could not even recognize individual letters, but he was determined to read again.

Two months after his stroke, still living at the rehab hospital, Howard had continuing difficulties recognizing places; he would get lost within the hospital three or four times a day and could not find his own room until he finally learned to recognize its floor "by the way the light filled the hall just opposite the elevator." He continued to have some object agnosia, too—even when he returned home after three months, he noted, "I kept finding cans of tuna in the dishwasher and jars of pencils in the freezer."

But with reading, Howard noted some signs of improvement: "the words no longer looked like they were written in an unfamiliar alphabet. The letters themselves looked like ordinary English letters, not the Serbo-Croatian I had imagined [after] my stroke."

There are two forms of alexia: a severe form which prevents even individual letters from being recognized and a milder form, in which letters can be recognized but only one by one, not simultaneously as words. Howard seemed to have moved, at this point, to the milder form—perhaps due to a partial recovery of the tissues affected by his stroke, or the brain's use (or perhaps even construction) of alternative pathways.[10]

Given this neurological improvement, he was able, with his therapists, to explore new ways of trying to read. He would slowly and

---

10. Brain damage from a stroke, a tumor, or a degenerative disease may produce a lasting alexia, but there can also be a transient alexia, due to a temporary disturbance in the brain's visual recognition systems, as can happen, for instance, with migraine. (This has been described by Fleishman et al. and Bigley and Sharp, among others.) I had such an experience driving to an appointment one morning, when I suddenly found myself unable to read the names of streets; they seemed to be written in a strange archaic script—Phoenician, perhaps—that I could not decipher. My first thought was of some external change. New York City is a popular location for filming, and the "altered" street signs, I presumed, were part of some elaborate cinematic setup. Then a sort of shimmering or scintillation around the letters gave me a clue—my alexia, I realized, was part of a migraine aura.

Alexia can also occur in conjunction with epilepsy. I recently saw a patient who described how reading (and only reading) triggers her seizures, but their first manifestation is an alexia. The words and letters before her suddenly become unintelligible, and she recognizes this as the prodrome of a seizure, which will follow within seconds. If she is alone, she will lie down and recite the alphabet to herself. On regaining consciousness after a seizure, she has expressive and receptive aphasia—an inability to speak or comprehend speech—for twenty minutes or so.

laboriously puzzle out words, letter by letter, forcing himself to decipher the names of streets and shops or the headlines of newspapers. "Familiar words," he said,

> including my own name, are unfamiliar blocks of type and have to be sounded out slowly. Each time a name recurs in an article or review, it hits me as unfamiliar on its last appearance as it does on the first.

Yet he persisted.

> Even though the reading was slow and difficult—frustrating as hell at times—I was still a reader. The blast to my brain could not make me otherwise. Reading was hard-wired into me. I could no more stop reading than I could stop my heart. . . . The idea of being cut off from Shakespeare and company left me weak. My life had been built on reading everything in sight.

Howard's reading grew somewhat easier with practice, though it might take him several seconds to make out a single word. "Words of different lengths," he observed, "like **cat, table** and **hippopotamus,** are processed in my head at a different rate. Each added letter adds more weight to the load that I am trying to lift." Scanning a page, reading in the usual sense, was still impossible, and "the whole process," he wrote, "was exhausting beyond belief." Sometimes, however, if he looked at a word, a couple of letters would suddenly jump out at him and be recognized—for example, the **bi** in the middle of his editor's name, though the letters before and after this remained unintelligible. He wondered whether such "chunking" was the way he had originally learned to read as a child, perhaps the way we all learn to read, before we go on to perceive words, even sentences, as a whole. (Pairs and perhaps clusters of letters are particularly important in the construction and reading of words, and whether reading is being learned for the first time or relearned after a stroke, there seems to be a natural progress from seeing single letters to seeing letter pairs or sequences. Dehaene and his colleagues suggest that there may be special "bigram" neurons in the brain devoted to this.)

"I can make myself see that certain letter groupings are indeed familiar words," Howard wrote to me, "but that comes only after I have stared at the page."

Becoming a fluent reader is a difficult and multileveled task; most children need years of practice and instruction to achieve this (though a few precocious ones may learn to read by themselves, and at an early age). In some ways, Howard had been reduced to the level of a child first learning his ABC's. But with a lifetime of experience as a reader, he could also bypass his disabilities to some extent, for his large vocabulary, his grammatical sense, and his command of literary and idiomatic English helped him to guess or infer words and even sentences from the slightest hint.

Whatever language a person is reading, the same area of inferotemporal cortex, the visual word form area, is activated. It makes relatively little difference whether the language uses an alphabet, like Greek or English, or ideograms, like Chinese.[11] This has been con-

---

11. There are some differences, however. As Maryanne Wolf points out, for example, "motoric memory areas are far more activated in reading Chinese than in reading other languages, because that is how Chinese symbols are learned by young readers—by writing, over and over." And the same reader may use somewhat different neural circuits for reading different languages.

One may sometimes find bilingual people who, following a stroke, lose the ability to read one language but not another. This has been especially studied in Japan, where there are two forms of written language in common use (often both forms are used in the same sentence). Kanji, which has a set of more than three thousand characters, was derived from Chinese ideograms. Kana, a syllabic system that, like an alphabet, can

firmed by lesion studies such as Déjerine's, and by imaging studies. And this idea is supported, too, by "positive" disorders—excesses or distortions of function produced by hyperactivity of the same area. The opposite of alexia, in this sense, is lexical or text hallucination, or phantom letters. People with disorders of the visual pathway (anywhere from the retina to the visual cortex) may be prone to visual hallucinations, and Dominic ffytche and his colleagues estimate that about a quarter of these patients who hallucinate see "text, isolated words, individual letters, numbers, or musical note hallucinations." Such lexical hallucinations, as ffytche and his colleagues have found, are associated with conspicuous activation of the left occipitotemporal region, especially the visual word form area—the same area that, if damaged, produces alexia.

So whether we are examining patients with alexia, patients with lexical hallucinations, or normal subjects reading, in any language, we are forced to the same conclusion: that there

---

represent any speech sound, has just forty-six symbols. Though kanji and kana are so different, both employ the visual word form area. Functional MRI studies by Nakayama and Dehaene, however, show subtle but significant differences in their representation within this area, and rare cases have been reported of alexia for kanji but not for kana, and vice versa.

exists, in every literate human being, an area in the dominant hemisphere—the language hemisphere—a neuronal system potentially available for the recognition of letters and words (and perhaps other forms of visual notation—mathematical or musical, for example).

This raises a deep problem: Why should all human beings have this built-in facility for reading, when writing is a relatively recent cultural invention?

Communication by the spoken word—and, therefore, its neural basis—has every mark of having evolved through the gradual processes of natural selection. The changing anatomy of the brain in prehistoric man has been worked out in some detail from endocranial casts and other fossil evidence, as have changes in the vocal tract. It is clear that the beginnings of speech go back hundreds of thousands of years. But this cannot be maintained in regard to reading, for writing emerged little more than five thousand years ago—far too recently to have occurred through evolution by natural selection. Though the visual word form area of the human brain appears so exquisitely tuned to the act of reading, it could not have evolved specifically for this purpose.

We might call this the Wallace problem, for Alfred Russel Wallace (who discovered natural selection independently of Darwin) became intensely concerned with the paradox of the human brain's many potential abilities—lexical, mathematical, and so on—abilities that would be of little use in a primitive or prehistoric society. While natural selection could explain the appearance of immediately useful abilities, he felt, it could not explain the existence of potential powers that might become manifest only with the development of an advanced culture hundreds of thousands of years in the future.

Unable to attribute these human potentials to any natural process, Wallace found himself constrained to invoke the supernatural: God, he believed, must have implanted them in the human psyche. There could hardly, from Wallace's perspective, be a better example of a divine gift—a unique new power, biding its time, **in posse**, waiting for the rise of a sufficiently advanced culture.[12]

---

12. Wallace expressed it as follows:

> Natural selection could only have endowed savage man with a brain a few degrees superior to that of an ape, whereas he actually possesses one very little inferior to that of a philosopher. . . . It seems as if the organ had been prepared in anticipation of the future progress in man, since it contains latent capacities which are useless to him in his earlier condition.

Darwin, understandably, was horrified by this idea and wrote to Wallace, "I hope you have not murdered too completely your own and my child." Darwin, for his part, had a much more open view of the process of natural selection and adaptation, foreseeing that biological structures might find uses very different from those for which they had originally evolved. (Stephen Jay Gould and Elisabeth Vrba called this sort of redeployment an "exaptation" rather than a direct adaptation.)[13]

How, then, did the visual word form area of the human brain arise? Does it exist in the brains of illiterate people? Does it have a precursor in the brains of other primates?

We are all faced with a world of sights and sounds and other stimuli, and our survival depends on making a rapid and accurate appraisal of these. Making sense of the world around us must be based on some sort of system, some swift and sure way of parsing the environment. Although seeing objects, defining them visually, seems to be instantaneous and innate, it represents a great perceptual achievement, one that requires a whole hierarchy of functions.

---

13. Gould provided a marvelous analysis of Wallace's thinking in his essay "Natural Selection and the Brain," reprinted in **The Panda's Thumb.**

We do not see objects as such; we see shapes, surfaces, contours, and boundaries, presenting themselves in different illumination or contexts, changing perspective with their movement or ours. From this complex, shifting visual chaos, we have to extract invariants that allow us to infer or hypothesize objecthood. It would be uneconomical to suppose that there are individual representations or engrams for each of the billions of objects around us. The power of combination must be called on; one needs a finite set or vocabulary of shapes that can be combined in an infinite number of ways, much as the twenty-six letters of the alphabet can be assembled (within certain rules and constraints) into as many words or sentences as a language ever needs.

There may be some objects that are recognized at birth, or soon after, like faces. But beyond this, the world of objects must be learned through experience and activity: looking, touching, handling, correlating the feel of objects with their appearance. Visual object recognition depends on the millions of neurons in the inferotemporal cortex, and neuronal function here is very plastic, open and highly responsive to experience and training, to education. Inferotemporal neurons evolved for general visual recognition, but they may be recruited for other purposes—most notably reading.

Such a redeployment of neurons is facilitated by the fact that all (natural) writing systems seem to share certain topological features with the environment, features which our brains evolved to decode. Mark Changizi, Shinsuke Shimojo, and their colleagues at Caltech examined more than a hundred ancient and modern writing systems, including alphabetic systems and Chinese ideograms, from a computational point of view. They have shown that all of them, while geometrically very different, share certain basic topological similarities. (This visual signature is not evident in artificial writing systems, such as shorthand, which are designed to emphasize speed more than visual recognition.) Changizi et al. have found similar topological invariants in a range of natural settings, and this has led them to hypothesize that the shapes of letters "have been selected to resemble the conglomerations of contours found in natural scenes, thereby tapping into our already-existing object recognition mechanisms."

Writing, a cultural tool, has evolved to make use of the inferotemporal neurons' preference for certain shapes. "Letter shape," Dehaene writes, "is not an arbitrary cultural choice. The brain constrains the design of an efficient writing system so severely that there is little room for cultural relativism. Our primate brain only accepts a limited set of written shapes."[14]

This is an elegant solution to the "Wallace problem"—indeed, it shows that there **is** no problem. The origin of writing and reading cannot be understood as a direct evolutionary adaptation. It is dependent on the plasticity of the brain, and the fact that even within the small span of a human lifetime, experience—

---

14. The earliest written languages used pictorial or iconic symbols, which became increasingly abstract and simplified. There were thousands of distinct hieroglyphs in Egypt and tens of thousands of ideograms in classical Chinese; reading (and writing) such a language demands a great deal of training and, presumably, the dedication of a larger portion of the visual cortex. This, Dehaene suggests, may be why most human languages have tended to favor alphabetic systems.

And yet there may be certain powers, certain qualities peculiar to ideograms. Jorge Luis Borges, who was well versed in Japanese poetry, spoke of the multiple connotations of kanji ideograms in an interview:

> The Japanese have achieved a wise ambiguity in their poetry. And that, I believe, is because of their particular form of writing itself, because of the possibilities that their ideograms present. Each one, according to its features, can have several connotations. Take, for example, the word "gold." This word represents or suggests autumn, the color of leaves, or the sunset because of its yellow color.

experiential selection—is as powerful an agent of change as natural selection. Natural selection, for Darwin, did not forbid cultural and individual developments on a timescale hundreds of thousands of times faster than evolutionary development—on the contrary, it prepared the ground for them. We are literate not by virtue of a divine intervention, but through a cultural invention and a cultural selection that makes a brilliant and creative new use of a preexisting neural proclivity.

While the visual word form area is crucial in the recognition of words and letters, many other areas of the brain are involved in "higher" levels of reading. This enabled Howard, for instance, to infer words from their context. Even now, nine years after his stroke, he is unable to recognize many simple words at a glance—but his writer's imagination does not just depend on reading.

While he was still in the rehab hospital, one of his therapists suggested that he keep a "memory book" to remind himself of appointments and to record his thoughts. As a lifelong keeper of journals, Howard was delighted by this idea. His new memory book proved to be an invaluable aid not only in stabilizing his still erratic memory but in reinforcing his identity as a writer:

> I knew I could no longer rely on the "sticking plaster" of memory. I could forget a word in the second part of what I was saying, even though I had already used the word a moment earlier. . . . I learned to write things down in the "memory book" [the moment I thought of them]. . . . The memory book gave a lift to my sense of being in the driver's seat of my life. [It] became my constant companion: part diary, part appointment book, part commonplace book. Hospitals, to a degree . . . breed a passive spirit; the memory book returned a piece of myself to me.

Keeping the memory book invited him, forced him, to write every day—not only at the level of forming legible words and sentences but at a much deeper creative level. His journal of hospital life, with its various routines and characters, began to stir his writer's imagination.

Occasionally, with unusual words or proper names, Howard might be unsure of their spelling—he could not "see" them in his mind's eye, imagine them, any more than he could perceive them when they were printed before him. Lacking this internal imagery, he had to employ other strategies for spelling. The simplest of these, he found, was to write a word in the air with his finger, letting a motor act take the place of a sensory one.

The great French neurologist Jean-Martin Charcot, in an 1883 lecture on a case of word blindness, describes a patient who, like Howard, has alexia sine agraphia. Charcot writes down the name of the hospital (which the patient himself has written earlier) and asks him to read it: "[The patient] is unable to do so at first; but he makes further efforts to do it and while he is accomplishing the task we notice that he traces, with the end of his right index finger, one of the letters which constitute the word, and with much trouble he says 'La Salpêtrière.' " When Charcot gives him the name of a street to read, the patient "traces with his finger in space the letters which compose the word, and after a moment or two says, 'It is the Rue d'Aboukir, the address of my friend.' "

Charcot's patient improved rapidly in "reading" by tracing letters in the air, and within three weeks, his reading speed had increased nearly sixfold. He said, "I can read printing less well than writing, because in writing it is easier for me to mentally reproduce the letter with my right hand, whereas it is more difficult to reproduce the printed characters." ("When reading printed matter," Charcot noted, "it is convenient for him to have a pen in his hand.") Concluding his lecture, Charcot emphasized, "Briefly put, one can say of him **that he reads only in the act of writing.**"

Increasingly and often unconsciously, then, Howard started to move his hands as he read, tracing the outlines of words and sentences still unintelligible to his eyes. And most remarkably, his tongue, too, began to move as he read, tracing the shapes of letters on his teeth or the roof of his mouth. This enabled him to read considerably faster (though it still might take him a month or more to read a book he could previously have read in an evening). Thus, by an extraordinary, metamodal, sensory-motor alchemy, Howard was replacing reading by a sort of writing. He was, in effect, reading with his tongue.[15]

---

15. Recently, while eating and talking, Howard bit the tip of his tongue by accident, and for a few days it was swollen and painful to move. He said, "It rendered me, for a day or so, illiterate once again."

The tongue, with its exquisite sensitivity, has an especially large motor and sensory representation in the brain. For this reason, it can be used for a sort of reading, as Howard does. Remarkably, it can also be used for sensory substitution devices that may enable blind people to "see" (see the chapter "The Mind's Eye").

· · ·

More than three months after his stroke, Howard returned from rehab to a home he did not entirely recognize:

> The house looked strange and familiar at the same time. . . . It was as though a movie set had been assembled from sketches of the real house and its rooms. Most peculiar was my office. I looked at my computer with a strange feeling. My whole office, where I had written several of my books, resembled a diorama in a museum. . . . On scribbled stick-on notes, my own handwriting looked strange, unfamiliar.

Would he ever be able to use this alien computer—once the main tool of his trade—again? With his son's help, and to his own surprise, he started to test out his old computer skills and soon felt them coming back. But writing something creative was another matter. And reading, even reading his own erratic handwriting, was still agonizingly slow and difficult. Furthermore, as he later wrote,

> I had been out of the world for months. I could no longer keep things straight in my head. What business did I have imagining

that I might go back to my old desk and begin again? I was clearly unfit for fiction. I turned off the computer and took a long walk.

Nonetheless, Howard had been, in a sense, staying in practice, writing every day, if only in his memory book. At first, he wrote,

> I had no thoughts of writing a book. That was not only well beyond my abilities, it was also beyond my imagination. But without my knowing it, another part of my brain was beginning to plot out a story. Images began popping into my head. Plots and plot twists began haunting my imagination. While I [had been] lying in my hospital bed . . . I was hard at work inventing story and characters and situations for the book I still didn't know I was writing.

He decided to write—if he could—a new novel, following his mother's old advice:

> **Write about what you know.** . . . What I knew about now was my illness. I knew the hospital routines and the people around me. I could do a book that described what it was like to be out of things, flat on my back for a time with nurses and doctors ordering and reordering my days.

He would reintroduce his alter ego, the detective Benny Cooperman, but it would be a Cooperman transformed: the great detective, waking in a hospital bed, finds himself not only alexic but amnesic as well. His powers of inference, however, are intact and enable him to stitch together disparate clues, to figure out how he landed in the hospital and what happened in the mysterious few days he can no longer remember.

Howard moved into high gear, typing for hours each day on his computer. Within a few weeks, his imagination and creative flow enabled him to produce a first draft. The problem now was how to correct and revise the draft, given his problems with short-term memory and his inability to read in the normal way. He employed many devices using his word processor—indenting certain paragraphs, marking passages with different font sizes—and after he had done as much as he could by himself, he got his editor to read the entire book aloud to him, so that he could engrave its overall structure in his memory and reorganize it in his mind. This painstaking process took many months of hard labor, but his abilities to remember and revise mentally, like Lilian Kallir's ability to arrange piano scores in her mind, steadily increased with practice.

His new novel (which he called **Memory Book**) was published in 2005, and this was followed in fairly rapid succession by another Benny Cooperman novel and, in 2007, a memoir, **The Man Who Forgot How to Read.** Howard Engel is still alexic, but he has found a way to remain a man of letters. That he was able to do so is a testament to many things: the dedication and skill of his therapists in rehab, his own determination to read again, and the adaptability of the human brain.

"The problems never went away," Howard writes, "but I became cleverer at solving them."

Face-Blind

IT IS WITH OUR FACES that we face the world, from the moment of birth to the moment of death. Our age and our sex are printed on our faces. Our emotions, the open and instinctive emotions which Darwin wrote about, as well as the hidden or repressed ones which Freud wrote about, are displayed on our faces, along with our thoughts and intentions. Though we may admire arms and legs, breasts and buttocks, it is the face, first and last, which is judged "beautiful" in an aesthetic sense, "fine" or "distinguished" in a moral or intellectual sense. And, crucially, it is by our faces that we can be recognized as individuals. Our faces bear the stamp of our experiences and character; at forty, it is said, a man has the face he deserves.

At two and a half months, babies respond to smiling faces by smiling back. "As the child smiles," Everett Ellinwood writes, "it usually engages the adult human to interact with him—to smile, to talk, to hold—in other words, to initiate the processes of socialization. . . . The reciprocal understanding mother-child relationship is possible only because of the continuing dialogue between faces." The face, psychoanalysts consider, is the first object to acquire visual meaning and significance. But are faces in a special category as far as the nervous system is concerned?

I have had difficulty recognizing faces for as long as I can remember. I did not think too much about this as a child, but by the time I was a teenager, in a new school, it was often a cause of embarrassment. My frequent inability to recognize schoolmates would cause them bewilderment and, sometimes, offense—it did not occur to them (why should it?) that I had a perceptual problem. I usually recognized close friends without much problem, especially my two best friends, Eric Korn and Jonathan Miller. But this was partly because I identified particular features: Eric had heavy eyebrows and thick spectacles, and Jonathan was tall and gangly, with a mop of red hair. Jonathan was a keen observer of postures, gestures, and facial expressions, and he seemingly never forgot a face. A decade later, when we were looking at old school photos, he could still recognize literally hundreds of our schoolmates, while I could not recognize a single one.

It was not just faces. When I went for a walk or a bicycle ride, I would have to follow exactly the same route, knowing that if I deviated from it even slightly, I would be instantly and hopelessly lost. I wanted to be adventurous, to go to exotic places—but I could do this only if I bicycled with a friend.

At the age of seventy-six, despite a lifetime of trying to compensate, I have no less trouble with faces and places. I am thrown particularly when I see people out of context, even if I have been with them five minutes before. This happened one morning just after my appointment with my psychiatrist (I had been seeing him twice weekly for several years at this point). A few minutes after I left his office, a soberly dressed man greeted me in the lobby of the building. I was puzzled as to why this stranger seemed to know me, until the doorman addressed him by name—it was, of course, my own analyst. (This failure to recognize him came up as a topic in our next session—I think he did not entirely believe me when I maintained that it had a neurological basis rather than a psychiatric one.)

A few months later, my nephew Jonathan Sacks came for a visit. We went out for a walk—

I lived in Mount Vernon, New York, at the time—and it started raining. "We had better get back," Jonathan said, but I couldn't find my house or my street. After two hours of walking around, in which we both got thoroughly soaked, I heard a shout. It was my landlord; he said he had seen me pass the house three or four times, apparently failing to recognize it.

In those years, I had to take the Boston Post Road to get from Mount Vernon to my hospital on Allerton Avenue in the Bronx. Though I took the same route twice a day for eight years, the road never became familiar to me, I never recognized the buildings on either side, and I would often turn the wrong way up the road, realizing it only when I came to one of two landmarks that were unmistakable, even for me: at one end, Allerton Avenue, which had a large sign, or, at the other, the Bronx River Parkway, which loomed over the Boston Post Road.

I had been working with my assistant, Kate, for about six years when we arranged to rendezvous in a midtown office for a meeting with my publisher. I arrived and announced myself to the receptionist, but failed to note that Kate had already arrived and was sitting in the waiting area. That is, I saw a young woman there, but did not realize it was her. After about five minutes, smiling, she said, "Hello, Oliver. I was wondering how long it would take you to recognize me."

Parties, even my own birthday parties, are a challenge. (More than once, Kate has asked my guests to wear name tags.) I have been accused of "absentmindedness," and no doubt this is true. But I think that a significant part of what is variously called my "shyness," my "reclusiveness," my "social ineptitude," my "eccentricity," even my "Asperger's syndrome," is a consequence and a misinterpretation of my difficulty recognizing faces.

My problem with recognizing faces extends not only to my nearest and dearest, but also to myself. Thus on several occasions I have apologized for almost bumping into a large bearded man, only to realize that the large bearded man was myself in a mirror. The opposite situation once occurred at a restaurant with tables outside. Sitting at one of these sidewalk tables, I turned to the restaurant window and began grooming my beard, as I often do. I then realized that what I had taken to be my reflection was not grooming himself but looking at me oddly. There was in fact a gray-bearded man on the other side of the window, who must have been wondering why I was preening myself in front of him.

Kate often cautions people in advance about

my little problem. She tells visitors, "Don't ask if he remembers you, because he will say no. Introduce yourself by name and tell him who you are." (And to me, she says, "Don't just say no—that's rude and will upset people. Say, 'I'm sorry, I am awful about recognizing people. I wouldn't recognize my own mother.' ")[1]

In 1988 I met Franco Magnani, the "memory artist," and over the next couple of years I spent weeks with him, talking about his paintings, his life, and even traveling to Italy with him to revisit the village where he grew up. When I finally submitted an article about him to **The New Yorker,** Robert Gottlieb, who was then the magazine's editor in chief, read the piece and said, "Very nice, fascinating—but what does he **look** like? Can you add some description?" I parried this awkward (and, to me, unanswerable) question by saying, "Who cares what he looks like? The piece is about his work."

---

1. This is an exaggeration—I had no trouble recognizing my parents or my brothers, though I was less adept with my huge extended family and completely lost, sometimes, when I saw photographs of them. I had dozens of aunts and uncles, and when I published my memoir **Uncle Tungsten,** I selected for the hardcover edition a photograph of another uncle, whom I mistakenly identified as Uncle Tungsten. This upset and bewildered his family, who said, "How could you make such a mistake? They look nothing like one another." (I corrected the error in the paperback edition.)

"Our readers will want to know," Bob said. "They need to picture him."

"I will have to ask Kate," I said. Bob gave me a peculiar look.

I assumed that I was just very bad at recognizing faces, as my friend Jonathan was very good—that this was within the limits of normal variation, and that he and I just stood at opposite ends of a spectrum. It was only when I went to Australia to visit my older brother Marcus, whom I had scarcely seen in thirty-five years, and discovered that he, too, had exactly the same difficulties recognizing faces and places that it dawned on me that this was something beyond normal variation, that we both had a specific trait, a so-called prosopagnosia, probably with a distinctive genetic basis.[2]

That there were others like me was brought home in various ways. The meeting of two peo-

---

2. Our other two brothers seemed to have normal powers of facial recognition. My father, a general practitioner, was immensely gregarious and knew hundreds of people, not to mention the thousands of patients in his practice. My mother, in contrast, was almost pathologically shy. She had a small circle of intimates—family and colleagues—and was very ill at ease in large gatherings. I cannot help wondering, in retrospect, if some of her "shyness" was due to a mild prosopagnosia.

ple with prosopagnosia, in particular, can be very challenging. A few years ago, I wrote to one of my colleagues to tell him that I admired his new book. His assistant then phoned Kate to arrange a meeting, and they settled on a weekend dinner at a restaurant in my neighborhood.

"There may be a problem," Kate said. "Dr. Sacks cannot recognize anyone."

"It's the same with Dr. W.," his assistant replied.

"And another thing," Kate added. "Dr. Sacks cannot find restaurants or other places; he gets lost very easily—he can't even recognize his own building sometimes."

"Yes, it's the same with Dr. W.," his assistant said.

Somehow, we did manage to meet and enjoyed dinner together. But I still have no idea what Dr. W. looks like, and he probably would not recognize me, either.

Although such examples may seem comical, they are sometimes quite devastating. People with very severe prosopagnosia may be unable to recognize their spouse, or to pick out their own child in a group of others.

Jane Goodall also has a certain degree of prosopagnosia. Her problems extend to recognizing chimpanzees as well as people—thus, she says, she is often unable to distinguish individual chimps by their faces. Once she knows a particular chimp well, she ceases to have difficulties; similarly, she has no problem with family and friends. But, she says, "I have huge problems with people with 'average' faces. . . . I have to search for a mole or something. I find it very embarrassing! I can be all day with someone and not know them the next day."

She adds that she, too, has difficulties in recognizing places: "I just don't know where I am until I am very familiar with the route. I have to turn and look at landmarks so I can find my way back. This was a problem in the forest, and I often got lost."

In 1985, I published a case history called "The Man Who Mistook His Wife for a Hat," about Dr. P., who had developed a very severe visual agnosia. He was not able to recognize faces or their expressions. Moreover, he could not identify or even categorize objects; thus, he was unable to recognize a glove, to recognize that it was an article of clothing, or that it resembled a hand. At one point he mistook his wife's head for his hat.

After Dr. P.'s story was published, I began to get letters from correspondents who would compare their difficulties in recognizing places and faces with his. In 1991, Anne F. wrote to me, describing her experiences:

> I believe that three people in my immediate family have visual agnosias: my father, a sister, and myself. We each have traits in common with your Dr. P., but, hopefully, not to the same degree. The most striking behavior we all share in common with Dr. P. is the prosopagnosia. My father, a man who has had a successful radio career here in Canada (his particular gift is an ability to mimic voices), was unable to recognize his wife in a recent photograph. At a wedding reception he asked a stranger to identify the man sitting next to his daughter (my husband of five years at the time).
>
> I have walked by my husband, while staring directly at his face, on several occasions without recognizing him. I have no difficulty recognizing him, however, in situations or places where I am expecting to see him. I am also able to recognize people immediately when they begin to speak, even if I've heard their voice only once in the past.
>
> Unlike Dr. P., I feel I can read people well on an emotional level. . . . I don't have the degree of agnosia for common objects that Dr. P. had. [However,] like Dr. P., I am totally incapable of establishing a topographical representation of space. . . . I have no memory

> for where I put things unless I verbally encode the location. Once an object leaves my hands, it drops off the edge of the world into a void.

While Anne F. seems to have prosopagnosia and topographical agnosia on a genetic or familial basis, others may develop this (or any other form of agnosia) in consequence of a stroke, a tumor, an infection, or an injury—or, like Dr. P., a degenerative disease such as Alzheimer's—that has damaged a particular part of the brain. Joan C., another correspondent, had an unusual history in this regard: she had developed a brain tumor in the right occipital lobe as an infant, and this was removed when she was two years old. It seems likely, though it is difficult to be certain, that her prosopagnosia was the result of either the tumor or the surgery. Her inability to recognize faces has often been misinterpreted by others. She notes, "I've been told that I'm rude, or a space cadet, or (according to a psychiatrist) suffering from a psychiatric disorder."

As I continued to receive more and more letters from people with prosopagnosia or topographical agnosia, it became clear to me that "my" visual problem was not uncommon and must affect many people around the world.

· · ·

Face recognition is crucially important for humans, and the vast majority of us are able to identify thousands of faces individually, or to easily pick out familiar faces in a crowd. A special expertise is needed to make such distinctions, and this expertise is nearly universal not only in humans but in other primates. How, then, do people with prosopagnosia manage?

In the last few decades, we have become very conscious of the brain's plasticity, how one part or system of the brain may take over the functions of a defective or damaged one. But this does not seem to occur with prosopagnosia or topographical agnosia; they are usually lifelong conditions that do not lessen as one grows older. People with prosopagnosia, therefore, need to be resourceful and inventive, need to find strategies, ways of circumventing their deficits: recognizing people by an unusual nose or beard, spectacles, or a certain sort of clothing.[3] Many

---

3. A most remarkable and creative reaction to face-blindness—the word "compensation" seems inadequate—is that of the artist Chuck Close, who is famous for his gigantic portraits of faces. Close himself has severe lifelong prosopagnosia. But this, he believes, played a crucial role in driving his unique artistic vision. He says, "I don't know who anyone is and have essentially no memory at all for people in real space, but when I

prosopagnosics recognize people by voice, posture, or gait; and, of course, context and expectation are paramount—one expects to see one's students at school, one's colleagues at the office, and so on. Such strategies, both conscious and unconscious, become so automatic that people with moderate prosopagnosia can remain unaware of how poor their facial recognition actually is, and are startled if it is revealed to them by testing (for example, with photographs that omit ancillary clues such as hair or eyeglasses).[4]

Thus, though I may be unable to recognize a particular face at a glance, I can recognize various things **about** a face: that there is a large nose, a pointed chin, tufted eyebrows, or protruding ears. Such features become identifying markers by which I recognize people. (I think, for similar reasons, I find it easier to recognize a caricature than a straightforward portrait or photograph.) I am reasonably good at judging age and gender, though I have made a few embarrassing blunders here. I am far better at

---

flatten them out in a photograph, I can commit that image to memory in a way; I have almost a kind of photographic memory for flat stuff."

4. It is similar with milder degrees of colorblindness or stereo blindness. People may be unaware of these "deficits," considering themselves normal, until the deficit is revealed through, for example, a routine eye examination or driver's license test.

recognizing people by the way they move, their "motor style." And even if I cannot recognize particular faces, I am sensitive to the beauty of faces, and to their expressions.[5]

I avoid conferences, parties, and large gath-

---

5. Once, as I was being interviewed on the radio about **The Man Who Mistook His Wife for a Hat**, a listener phoned in and said, "I can't recognize my wife, either." (This, he added, was because he had developed a brain tumor.) I arranged to see Lester C. and find out more about his experiences.

While Lester had found various strategies for recognizing people, he told me, he was distressed by his inability to appreciate the beauty of faces. He had had "a great eye for the girls," he said, before the tumor. Now he had to judge beauty indirectly, by taking seven criteria (color of eyes, shape of nose, symmetry, etc.) and rating each of these on a scale of one to ten. This way he could construct a "mental histogram," as he put it, for beauty. But he soon found that such histograms did not work and were sometimes ludicrously at odds with a direct or intuitive judgment of beauty such as he had once had.

Most people with prosopagnosia remain sensitive to facial expressions, seeing at a glance whether someone looks happy or sad, friendly or hostile, even though the faces themselves may be unidentifiable. The reverse also occurs: Antonio Damasio has described how people with damage to the amygdala (a part of the brain crucial to the perception and feeling of emotion) may have difficulty "reading" faces, judging their emotional expressions, even though they recognize faces normally. This may also be the case with some autistic people. Temple Grandin, who has Asperger's syndrome, says, "I can recognize major expressions on a person's face, but I do not pick up subtle cues. I did not know that people had little eye signals until I read about them in Simon Baron-Cohen's book **Mindblindness** when I was fifty." (Though Temple is a "visual thinker" and can

erings as much as I can, knowing that they will lead to anxiety and embarrassing situations—not only failing to recognize people I know well, but greeting strangers as old friends. (Like many prosopagnosics, I avoid greeting people by name, lest I use the wrong one, and I depend on others to save me from egregious social blunders.)

I am much better at recognizing my neighbors' dogs (they have characteristic shapes and colors) than my neighbors themselves. Thus when I see a youngish woman with a Rhodesian ridgeback hound, I realize that she lives in the apartment next to mine. If I see an older lady with a friendly golden retriever, I know this is someone from down the block. But if I should pass either woman on the street without her dog, she might as well be a complete stranger.

The idea that "the mind"—an immaterial, airy thing—could be embodied in a

---

easily visualize complicated engineering problems, she seems to be no better or worse than average at recognizing faces.)

Difficulty making social contact with others can also be a central problem in schizophrenia, and Yong-Wook Shin et al. have obtained preliminary results suggesting that schizophrenic people have difficulty not only in reading facial expressions but in face recognition, too.

lump of flesh—the brain—was intolerable to seventeenth-century religious thinking; hence the dualism of Descartes and others. But physicians, observing the effects of strokes and other brain injuries, had long had reason to suspect that the functions of the mind and brain were linked. Toward the end of the eighteenth century, the anatomist Franz Joseph Gall proposed that all mental functions must arise from the brain—not from the "soul," as many people imagined, or from the heart or the liver. Instead, he envisioned within the brain a collection of twenty-seven "organs," each responsible for a different moral or mental faculty. Such faculties, for Gall, included what we would now call perceptual functions, such as the sensation of color or sound; cognitive faculties, like memory, mechanical aptitude, or speech and language; and even "moral" traits such as friendship, benevolence, or pride. For these heretical ideas, he was exiled from Vienna and wound up eventually in revolutionary France, where he hoped a more scientific approach might be embraced.[6]

---

6. Determined to provide some objective correlate, Gall went further, trying to measure and correlate personality and moral faculties of individuals with the shapes and bumps of their skulls, using a method he called "cranioscopy." One of his students, Johann Spurzheim, went on to popularize this idea as

The physiologist Jean-Pierre Flourens decided to investigate Gall's theory by removing slices of the brain in living animals, chiefly pigeons. But he could not find any evidence to correlate specific areas of the cortex with specific faculties (perhaps because one needs very delicate and discrete ablations to do so, especially in the tiny pigeon cortex). So Flourens believed that the cognitive impairments his pigeons exhibited as he removed more pieces of cortex reflected only the amount of cortex removed, not its location, and what applied to birds, he felt, probably also applied to human beings. The cortex, he concluded, was equipotential, as homogeneous and undifferentiated as the liver. "The brain," Flourens said, only half jesting, "secretes thought as the liver secretes bile."

Flourens's notion of an equipotential cortex dominated thought until the studies of Paul Broca in the 1860s. Broca performed autopsies on many patients with expressive aphasia, all of whom, he showed, had damage limited to the frontal lobes on the left side. In 1865, he

---

"phrenology," a pseudoscience that gained much attention in the early nineteenth century and influenced Lombroso's theories of criminal physiognomy. Spurzheim and Lombroso's work has long been discredited, but Gall's idea of localization in the brain had a lasting impact.

was able to say, famously, "We speak with our left hemisphere," and the notion of a homogeneous and undifferentiated brain, it seemed, was laid to rest.

Broca felt that he had located a "motor center for words" in a particular part of the left frontal lobe, an area we now call Broca's area.[7] This seemed to promise a new sort of localization, a genuine correlation of neurological and cognitive functions with specific centers in the brain. Neurology moved confidently ahead, identifying "centers" of every sort: Broca's motor center for words was followed by Wernicke's auditory center for words, and Déjerine's visual center for words, all in the left hemisphere, the language hemisphere, and a center for visual recognition in the right hemisphere.

---

7. In 1869, Hughlings Jackson debated this issue with Broca, insisting that "to locate the damage which destroys speech and to locate speech are two different things." Jackson, it was generally thought, lost this debate, but he was not the only one with reservations. Freud, in his 1891 book **On Aphasia,** suggested that the use of language demanded many interconnected areas of the brain, and that Broca's area was only one node in a vast cerebral network. The neurologist Henry Head, in his monumental 1926 treatise **Aphasia and Kindred Disorders of Speech,** inveighed against "the diagram-makers," as he called the aphasiologists of the nineteenth century. Head argued, as Hughlings Jackson and Freud had, for a much more holistic view of speech.

But while visual agnosia of a general sort was recognized in the 1890s, there was little idea that there could be agnosia for particular visual categories like faces or places—even though major figures like Hughlings Jackson and Charcot had already described specific agnosias for faces and places following damage to the posterior areas of the right hemisphere. In 1872, Jackson described a man who, following a stroke in this area, lost his ability "to recognize places and persons. At one time he did not know his wife . . . and having wandered from home was unable to find his way back." Charcot, in 1883, provided an account of a patient who had enjoyed exceptional powers of visual imagery and memory, but lost these suddenly. Charcot describes how this man "cannot even recall his own face. Recently in a public gallery his path seemed to be stopped by a person to whom he was about to offer his excuses, but it was merely his own image reflected in a glass."

Still, even by the middle of the twentieth century, many neurologists doubted whether the brain had category-specific recognition areas. This may have played a part in delaying the recognition of face-blindness, despite the evidence from clinical cases.

In 1947, Joachim Bodamer, a German neurologist, described three patients who were unable to recognize faces but had no other difficulties with recognition. It seemed to Bodamer that this highly selective form of agnosia needed a special name—it was he who coined the term "prosopagnosia"—and that such a specific loss must imply that there was a discrete area in the brain specialized for face recognition. This has been a matter of dispute ever since: is there a special system dedicated only to face recognition, or is face recognition simply one function of a more general visual recognition system? Macdonald Critchley, writing in 1953, was highly critical of Bodamer's article and of the very idea of face-blindness. "It seems scarcely credible," he wrote, "that human faces should occupy a perceptual category which is different from all other objects in space, animate and inanimate. Can there be any attribute of size, shape, colouring or motility which distinguishes a human face from other objects in such a way as to preclude identification?"

But in 1955, the English neurologist Christopher Pallis published a beautifully detailed and documented study of his patient A.H., a mining engineer at a Welsh colliery who had kept a journal and was able to give Pallis an articulate and insightful description of his experiences. One night in June of 1953, A.H. apparently suffered a stroke. He "suddenly felt unwell after a couple of drinks at his club." He appeared to be confused and was taken home to bed, where he slept poorly. Getting up the following morning, he found his visual world completely transformed, as he reported to Pallis:

> I got out of my bed. My mind was clear but I could not recognize the bedroom. I went to the toilet. I had difficulty finding my way and recognizing the place. Turning round to go back to bed I found I couldn't recognize the room, which was a strange place to me.
>
> I could not see colour, only being able to distinguish light objects from dark ones. Then I found out all faces were alike. I couldn't tell the difference between my wife and my daughters. Later I had to wait for my wife or mother to speak before recognizing them. My mother is 80 years old.
>
> I can see the eyes, nose, and mouth quite clearly but they just don't add up. They all seem chalked in, like on a blackboard.

His difficulty was not limited to recognizing people in real life:

> I cannot recognize people in photographs, not even myself. At the club I saw someone strange staring at me and asked the steward who it was. You'll laugh at me. I'd been look-

ing at myself in a mirror. . . . I later went to London and visited several cinemas and theatres. I couldn't make head or tail of the plots. I never knew who was who. . . . I bought some copies of **Men Only** and **London Opinion.** I couldn't enjoy the usual pictures. I could work out what was what by accessory details, but it's no fun that way. You've got to take it in at a glance.

A.H. had other visual problems: a small defect in one corner of his visual fields, transient difficulty with reading, a total inability to perceive color, and difficulty identifying places. (He had initially had some odd sensations on the left side, too—a "heaviness" of the left hand and a "stinging" feeling in his left index finger and the left corner of his mouth.) But he had no object agnosia: he was able to sort out geometrical figures, to draw complex objects, to assemble jigsaw puzzles and play chess.

Since Pallis's time, a number of patients with prosopagnosia have come to autopsy. Here the data are clear: virtually all patients who acquire prosopagnosia, irrespective of the cause, have lesions in the right visual association cortex, in particular on the underside of the occipitotemporal cortex; there is nearly always damage in a structure called the fusiform gyrus. These autopsy results gained additional support in the 1980s, when it became possible to visualize the brains of living patients by using CT scans and MRIs—here, too, prosopagnosic patients showed lesions in what came to be called the "fusiform face area." (Abnormal activity in the fusiform face area has also been correlated with hallucination of faces, as Dominic ffytche and his colleagues have shown.)

In the 1990s, such lesion studies were complemented by functional imaging—visualizing the brains of people with fMRIs as they looked at pictures of faces, places, and objects. These functional studies demonstrated that looking at faces activated the fusiform face area much more strongly than looking at other test images.

That individual neurons in this area could show preferences was first demonstrated in 1969 by Charles Gross and his colleagues, using electrodes in the inferotemporal cortex of macaques. Gross found cells that responded dramatically to the sight of a monkey's paw—but also, less strongly, to a variety of other stimuli, including a human hand. Subsequently, he found cells with a relative preference for faces.[8]

---

8. Much that we now take for granted in neuroscience was very unclear when Gross began this work. Even in the late 1960s, it was widely believed that the visual cortex did not extend far

At this purely visual level, faces are distinguished as configurations, in part by detecting the geometrical relationships between eyes, nose, mouth, and other features (as Freiwald, Tsao, and Livingstone have established).[9] But there is no preference at this level for individual faces; indeed, generic or cartoon faces can elicit the same responses as real ones.

Recognition of particular faces or objects is only achieved at a higher cortical level, in the multimodal area of the medial temporal lobe, which has rich reciprocal connections not only to the fusiform face area but to other areas subserving sensory association, emotion,

---

beyond its main locus in the occipital lobes (as we now know it does). That the representation and recognition of specific categories of objects—faces, hands, etc.—might rely on individual neurons or clusters of neurons was considered improbable, even absurd; the idea was good-humoredly mocked by Jerome Lettvin in his famous comments about "grandmother cells." Very little attention, therefore, was paid to Gross's early findings, and it was not until the 1980s that they were confirmed and amplified by other researchers.

9. Different inferotemporal cells, they write, are "selective for different face parts and interactions between parts, and even the same cell can respond maximally to different combinations of face parts. Thus, there is no single blueprint for detecting the form of a face. . . . This diversity of feature tuning provides the brain with a rich vocabulary to describe faces and shows how a high-dimensional parameter space may be encoded even in a small region of [the inferotemporal cortex]."

and memory. Christof Koch, Itzhak Fried, and their colleagues have shown that cells in the multimodal medial temporal lobe area show remarkable specificity, responding only, for example, to images of Bill Clinton, or spiders, or the Empire State Building, or cartoons from **The Simpsons.** Specific neural units may also respond to hearing or reading the name of the person or object; thus in one patient, a set of neurons responded strongly to pictures of the Sydney Opera House and also to the letter string "Sydney Opera," though not to the names of other landmarks, such as "Eiffel Tower."[10]

Neurons in the medial temporal lobe are capable of encoding representations of individual faces, landmarks, or objects so that they can be easily recognized in a changing environment. Such representations can be constructed rapidly, within less than a day or two after exposure to an unfamiliar individual.

Although such studies involve electrode recordings from single neurons, each of these cells is connected to thousands of other neurons, each of which in turn is connected to thousands more. (Some single cells, moreover,

---

10. Koch, Fried, and their colleagues have published many papers on their work; those most relevant here include Quian Quiroga et al., 2005 and 2009.

may respond to more than one individual or object.) So a single cell's response really represents the apex of an immense computational pyramid, perhaps drawing on direct or indirect inputs from the visual, auditory, or tactile cortex, text-recognition areas, memory and emotional areas, and so on.

In humans, some ability to recognize faces is present at birth or soon after. By six months, as Olivier Pascalis and his colleagues have shown in one study, babies are able to recognize a broad variety of individual faces, including those of another species (in this study, pictures of monkeys were used). By nine months, though, the babies became less adept at recognizing monkey faces unless they had received continuing exposure to them. As early as three months, infants are learning to narrow their model of "faces" to those they are frequently exposed to. The implications of this work for humans are profound. To a Chinese baby brought up in his own ethnic environment, Caucasian faces may all, relatively, "look the same," and vice versa.[11] One prosopagnosic acquaintance, born and raised in China, went to Oxford as a student and has lived for decades in the United States.

---

11. Yoichi Sugita points out, however, that this narrowing is easily reversible, at least in childhood, by experience.

Nonetheless, he tells me, "European faces are the most difficult—they all look the same to me." It seems that there is an innate and presumably genetically determined ability to recognize faces, and this capacity gets focused in the first year or two, so that we become especially good at recognizing the sorts of faces we are likely to encounter. Our "face cells," already present at birth, need experience to develop fully.

It is similar with many other capacities, from stereo vision to linguistic power: some predisposition or potential is built in genetically but requires stimulation, practice, environmental richness, and nourishment if it is to develop fully. Natural selection may bring about the initial predisposition, but experience and experiential selection are needed to bring our cognitive and perceptual capacities to their full realization.

The fact that many (though not all) people with prosopagnosia also have difficulty with recognizing places has suggested to some researchers that face and place recognition are mediated by distinct yet adjacent areas. Others believe that both are mediated by a single zone which is perhaps more oriented to faces at one end and to places at the other.

The neuropsychologist Elkhonon Goldberg, however, questions the whole notion of discrete, hardwired centers or modules with fixed functions in the cerebral cortex. He feels that at higher cortical levels there may be much more in the way of gradients, where areas whose function is developed by experience and training overlap or grade into one another. In his book **The New Executive Brain**, he speculates that a gradiential principle constitutes an evolutionary alternative to a modular one, permitting a degree of flexibility and plasticity impossible for a brain organized in a purely modular fashion.

While modularity, he argues, may be characteristic of the thalamus—an assemblage of nuclei with fixed functions, fixed inputs and outputs—a gradiential organization is more characteristic of the cerebral cortex, and becomes more and more prominent as one ascends from primary sensory cortex to association cortex, to the highest level of all, the frontal cortex. Modularity and gradients may thus coexist and complement one another.

People with prosopagnosia, even if their chief complaint is of face-blindness, often have difficulty recognizing other specific things. Orrin Devinsky and Martha Farah have remarked that some prosopagnosics are unable to distinguish an apple from a pear, say, or a pigeon from a raven, although they can correctly recognize the general category of "fruit" or "bird." Joan C. described a similar problem: "I don't recognize handwriting in the same way that I don't recognize faces. That is, I might be able to identify a sample of handwriting by recognizing some salient feature or by seeing it in context, but otherwise, forget it. I've even failed to recognize my own handwriting."

Some researchers have proposed that prosopagnosia is not purely a problem with face-blindness, but one aspect of a more general difficulty in distinguishing the individuals in any class, whether the class is of faces, cars, birds, or anything else.

Isabel Gauthier and her colleagues at Vanderbilt tested a group of car experts and a group of expert birders, comparing them to a group of normal subjects. The fusiform face area, they found, was activated when all of the groups looked at pictures of faces. But it was also activated in the car experts when they were asked to identify particular cars, and in the birders when they were asked to identify particular birds. The fusiform face area is primarily tuned for facial recognition, but some of it, it seems, can be trained to distinguish individual items of other sorts. (If, then, an expert bird spotter or car buff is unlucky enough to acquire prosopagnosia, he will also, we might suspect, lose his facility for identifying birds or cars.)

The brain is more than an assemblage of autonomous modules, each crucial for a specific mental function. Every one of these functionally specialized areas must interact with dozens or hundreds of others, their total integration creating something like a vastly complicated orchestra with thousands of instruments, an orchestra that conducts itself, with an ever-changing score and repertoire. The fusiform face area does not work in isolation; it is a vital node in a cognitive network that stretches from the occipital cortex to the prefrontal area. Face-blindness may occur even with an intact fusiform face area, if the lower occipital face areas are damaged. And people with moderate prosopagnosia, like Jane Goodall or myself, can, after repeated exposure, learn to identify those we know best. Perhaps this is because we are using slightly different pathways to do so, or perhaps, with training, we can make better use of our relatively weak fusiform face areas.

Above all, the recognition of faces depends not only on the ability to parse the visual aspects of a face—its particular features and their overall configuration—and compare it to others, but the ability to summon the memories, experiences, and feelings associated with that face. The recognition of specific places or faces, as Pallis emphasized, goes with a particular feeling, a sense of association and meaning. While purely visual recognition of faces is mediated by the fusiform face area and its connections, emotional familiarity is mediated at a higher, multimodal level, where there are intimate connections with the hippocampi and amygdala, areas dedicated to memory and emotion. Thus A.H., after his stroke, lost not only his ability to identify faces but this sense of familiarity; every face and place appeared new to him and continued to do so even if seen again and again.

Recognition is based on knowledge; familiarity is based on feeling; but neither entails the other. The two have different neural bases and can be dissociated; thus, although both are lost in tandem with prosopagnosia, one can have familiarity without recognition or recognition without familiarity in other conditions. The former occurs in déjà vu and also in the "hyperfamiliarity" for faces described by Devinsky. Here a patient may find that everyone on the bus or on the street looks "familiar"—he may go up to them and address them as old friends, even while realizing that he cannot possibly know them all. My father was always very sociable and could recognize hundreds or even thousands of people, but his feeling of "knowing" people became exaggerated, perhaps pathological, as he moved into his nineties. He often attended concerts at the Wigmore Hall in London, and there, during the intermissions, he would accost everyone in sight, saying, "Don't I know you?"

The opposite occurs in patients with Capgras syndrome, for whom people's faces, though recognized, no longer generate a sense of emotional familiarity. Since a husband or wife or child does not convey that special warm feeling of familiarity, the Capgras patient will argue, they cannot be the real thing—they **must** be clever impostors, counterfeits. People with prosopagnosia have insight; they realize that their problems with recognition come from their own brains. People with Capgras syndrome, in contrast, remain immovable in their conviction that they are perfectly normal and it is the other person who is profoundly, even uncannily wrong.

People with acquired prosopagnosia, like A.H. or Dr. P., are relatively rare—most neurologists are likely to encounter such a patient once or twice in their career, if at all. Congenital prosopagnosia (or, as it is sometimes called, "developmental" prosopagnosia), such as I have, is much commoner, yet remains completely unrecognized by most neurologists. Heather Sellers, a lifelong prosopagnosic, wrote about this in a 2007 autobiographical essay: "I couldn't recognize my husband's children. . . . I hugged the wrong man in the grocery, thinking it was [my husband]. . . . My colleagues remained unidentifiable after a decade. . . . I kept introducing myself to neighbors." When she consulted two separate neurologists for her problem, they both said that they had never seen it before, and it was "very rare."[12]

---

12. Despite its unfamiliarity to modern physicians, congenital prosopagnosia entered the medical literature as early as 1844, when A. L. Wigan, an English doctor, described one of his patients:

> A gentleman of middle age . . . lamented to me his utter inability to remember faces. He would converse with a person for an hour, but after an interval of a day could not recognise him again. Even friends, with whom he had been engaged in business transactions, he was unconscious of ever having seen. Being in an occupation in which it was essential to cultivate the good-will of the public, his life was made perfectly miserable by this unfortunate defect, and his time was passed in offending and apologizing. He was quite incapable of making a mental picture of anything, and it was not till he heard the voice, that he could recognise men with whom he had constant intercourse. . . . I endeavoured in vain

One eminent neurologist who has written about visual agnosia confessed to me that he had not even heard of congenital prosopagnosia until very recently. This, however, is not entirely surprising, for people with congenital prosopagnosia do not generally consult neurologists about their "problem," any more than someone with lifelong colorblindness would complain about it to an eye doctor. It is just the way they are.

But Ken Nakayama at Harvard, who investigates visual perception, has long suspected that prosopagnosia is relatively common but underreported. In 1999, he and his colleague Brad Duchaine, at University College London, began using the internet to seek subjects with face-blindness, and they received a startling response. They are now investigating several thousand people with lifelong prosopagnosia ranging from mild to cripplingly severe.[13]

While people with lifelong prosopagnosia do not have gross lesions in the brain, a re-

---

to convince him that an acknowledgment of the defect would be the best means of removing the unfortunate effect it had produced in alienating friends. He was quite determined to conceal it, if possible, and it was impossible to convince him that it did not depend solely on the eyes.

13. Information is available at their website, www.faceblind.org.

cent study by Lucia Garrido and her colleagues showed that they do have subtle but distinct changes in the brain's face-recognition areas. The condition also tends to be familial: Duchaine, Nakayama, and their colleagues have described one family in which ten members—both parents and seven of their eight children (the eighth could not be tested), as well as a maternal uncle—have it. Clearly there are strong genetic determinants at work here.

Nakayama and Duchaine have explored the neural basis of face and place recognition, generating new knowledge and insights at every level from the genetic to the cortical. They have also studied the psychological effects and social consequences of developmental prosopagnosia and topographical agnosia—the special problems these conditions can create for an individual in a complex social and urban culture.

The range seems to extend in a positive direction, too. Russell, Duchaine, and Nakayama have described "super-recognizers," people with extraordinarily good face-recognition abilities, including some who seem to have indelible memories of virtually every face they have ever seen. Alexandra Lynch, one of my correspondents, described her own uncanny ability to recognize people:

It happened again yesterday. I was on my way down into the subway in Soho when I identified someone fifteen feet ahead of me (back turned, talking intimately with his friend) as a man I knew, or had seen before. In this case, it was Mac, who used to be a family friend's art dealer. I had last seen him (briefly) two years earlier, at an opening in midtown. I'm not sure I've ever spoken with him beyond an introduction a good ten years ago.

This is an integral part of my life—I catch a passing glimpse of someone and, with no real effort, **flash**, place the face—yes, that's the girl who served us wine at an East Village bar last year (again, in a totally different neighborhood, and at night not during the day). It is true that I'm a big fan of people, of humanity and diversity . . . but to my knowledge I make no effort to record the physical traits of ice cream servers, shoe salesmen and friends of friends of friends. Even a slim wedge of face, or the way someone walks two blocks away at dusk, can trigger my mind to zero in on a match.

The super-recognizers, Russell et al. write, "are about as good as many [lifelong] prosopagnosics are bad"—that is, they are about two or three standard deviations above average, while the most severe prosopagnosics have face-recognizing abilities two or three standard deviations below average. Thus the difference between the best face recognizers and the worst among us is comparable to that between people with an IQ of 150 and an IQ of 50, with others filling every level in between. As with any bell curve, the vast majority of people are somewhere in the middle.

Severe congenital prosopagnosia is estimated to affect at least 2 percent of the population—six million people in the United States alone. (A much higher percentage, perhaps 10 percent, are markedly below average in face identification but not cripplingly face-blind.) For these people, who have difficulty recognizing their husbands, wives, children, teachers, and colleagues, there is still no official recognition or public understanding.

This is in marked contrast to the situation with another neurological minority, the 5 to 10 percent of the population with dyslexia. Teachers and others are more and more aware of the special difficulties and often special gifts which dyslexic children may have, and are starting to provide educational strategies and resources for them.

But for now, people with varying degrees of face-blindness must rely on their own ingenuity and strategies, starting with educating others about their unusual, but not rare, condition. Increasingly, prosopagnosia is the subject of books, websites, and support groups, where people with face-blindness or topographical agnosia are able to share experiences and, no less important, strategies for recognizing faces and places when the usual "automatic" mechanisms are compromised.

Ken Nakayama, who is doing so much to further the scientific understanding of prosopagnosia, also has a personal acquaintance with the subject, and posts this notice in his office and on his website:

> Recent eye problems and mild prosopagnosia have made it harder for me to recognize people I should know. Please help by giving your name if we meet. Many thanks.

Stereo Sue

When Galen, in the second century, and Leonardo, thirteen centuries later, observed that the images received by the two eyes were slightly different, neither of them appreciated the full significance of these differences. It was not until the early 1830s that Charles Wheatstone, a young physicist, began to suspect that even though the brain somehow fused these images automatically and unconsciously, the disparities between the two retinal images were in fact crucial to the brain's mysterious ability to generate a sensation of depth.

Wheatstone confirmed the truth of his conjecture by an experimental method as simple as it was brilliant. He made pairs of drawings of a solid object as seen from the slightly different perspectives of the two eyes and then designed an instrument that used mirrors to insure that each eye saw only its own drawing. He called the device a stereoscope, from the Greek for "solid vision." If one looked into the stereoscope, the two flat drawings would fuse to produce a single three-dimensional drawing poised in space.

(One does not need a stereoscope to see stereo depth; it is relatively easy for most people to learn how to "free-fuse" such drawings, simply by diverging or converging the eyes. So it is strange that stereopsis was not discovered centuries before: Euclid or Archimedes could have drawn stereo diagrams in the sand, as David Hubel has remarked, and discovered stereopsis in the third century b.c. But they did not, as far as we know.)

Photography was invented only months after Wheatstone's 1838 article describing his stereoscope, and stereo photographs quickly became popular.[1] Queen Victoria herself was presented with a stereoscope after admiring

---

1. Wheatstone's name is more commonly associated with the invention of the Wheatstone bridge, an instrument used to measure electrical resistance. But like several other eminent nineteenth-century scientists, Wheatstone was also deeply interested in the physical basis of perception. All of these "natural philosophers" (we would now call them physicists), using ingenious experiments, contributed to our understanding of how the eye and brain construct our perceptions of depth and movement and color, as they also contributed to the technological development of stereo, cinematic, and color photography.

one in the Great Exhibition at the Crystal Palace, and soon no Victorian drawing room was complete without one. With the development of smaller, cheaper stereoscopes, easier photographic printing, and even stereo parlors, there were few people in Europe or America who did not have access to stereo viewers by the end of the nineteenth century.

With stereo photographs, viewers could see the monuments of Paris and London or great sights of nature like Niagara Falls or the Alps in all their majesty and depth, with an uncanny verisimilitude that made them feel as if they were hovering over the actual scenes.[2]

In 1861, Oliver Wendell Holmes (who in-

Michael Faraday, in addition to his electromagnetic studies, played a part in devising zoetrope-like instruments that presented a series of still drawings to the eyes in rapid succession, demonstrating that at a critical rate these could be fused by the brain to create a sensation of motion.

James Clerk Maxwell was intrigued by Thomas Young's hypothesis that there were three—and only three—distinct types of color receptors in the retina, each responsive to light of a certain wavelength (roughly corresponding to red, green, and blue). He devised an elegant test of this by photographing a colored bow through red, green, and violet filters and then projecting the three photographs through their corresponding filters. When the three monochromatic images were perfectly superimposed, the picture burst into full color.

2. By the mid-1850s, a subspecialty of stereo photography, stereo pornography, was already well established, though this was

vented the popular handheld Holmes Stereo Viewer), in one of several **Atlantic Monthly** articles on stereoscopes, remarked on the special pleasure people seemed to derive from this magical illusion of depth:

> The shutting out of surrounding objects, and the concentration of the whole attention . . . produces a dream-like exaltation . . . in which we seem to leave the body behind us and sail into one strange scene after another, like disembodied spirits.

There are, of course, many other ways of judging depth besides stereo vision: the occlusion of distant objects by closer objects, perspective (the fact that parallel lines converge as they recede, and that distant objects appear smaller), shading (which delineates the shape of objects), "aerial" perspective (the blurring and blueing of more distant objects by the intervening air), and, most important, motion parallax—the changing appearance of spatial relationships as we move around in the world. All these cues, acting together, can give a sense of reality and space and depth. But the only

---

of a rather static type, because the photographic processes used at the time required lengthy exposures.

way to actually **perceive** depth—to see it rather than judge it—is with binocular stereoscopy.[3]

In my boyhood home, in London during the 1930s, we had two stereoscopes: a large, old-fashioned wooden one, which took glass slides, and a smaller handheld one, which took cardboard stereo photographs. We also had books of bicolor anaglyphs—stereo photographs printed in red and green, which had to be viewed with a pair of glasses with one red and one green lens, which effectively restricted each eye to seeing only one of the images.

So when, at the age of ten, I developed a passion for photography, I wanted, of course, to make my own pairs of stereo photos. This was easy to do, by moving the camera horizontally about two and a half inches between exposures, mimicking the distance between the two eyes. (I did not yet have a double-lens stereo camera, which would take simultaneous stereo pairs.)

After reading how Wheatstone explored ste-

---

3. There is one situation, as I learned by painful experience, when two eyes do not help. When I was growing up, we always had a clothesline strung across the garden, and since it traversed the entire visual field horizontally, it appeared exactly the same to both eyes, and I could never judge how far away it was. I had to approach it cautiously, since it was strung rather low, at about the height of my neck. Sometimes, forgetting this, I would run straight into it, almost garroting myself.

reoscopic effects by exaggerating or reversing the disparity between the two images, I began experimenting with this, too. I started taking pictures with greater and greater separations between them, and then I made a hyperstereoscope, using a cardboard tube about a yard long with four little mirrors. With this, I could turn myself, in effect, into a creature with eyes a yard apart. I could look through the hyperstereoscope at a very distant object, like the dome of St. Paul's Cathedral, which normally appeared as a flat semicircle on the horizon, and see it in its full rotundity, projecting towards me. I also experimented with making a "pseudoscope," which transposed the views of the two eyes to reverse the stereo effect to some extent, making distant objects appear closer than near ones and even turning faces into hollow masks. This, of course, contradicted common sense, as well as all the other depth cues of perspective and occlusion—sometimes the images would rapidly shift back and forth from convex to concave, a bizarre and disorienting experience as the brain struggled to reconcile two rival hypotheses.[4]

---

4. Richard Gregory, who studied visual illusions for many years, insisted that perceptions were, in fact, perceptual hypotheses (as, in the 1860s, Hermann von Helmholtz called them "unconscious inferences"). Gregory was a stereo enthusiast—he often sent his friends stereoscopic Christmas cards—but when

After the Second World War, new techniques and forms of stereoscopy became popular. The View-Master, a little stereoscope made of plastic, took reels of tiny Kodachrome transparencies that one flicked through by pressing a lever. I fell in love with faraway America at this time, partly through View-Master reels of the grand scenery of the American West and Southwest.

One could also get Polaroid Vectographs, in which the stereo images were polarized at right angles to each other; these were viewed through a special pair of Polaroid glasses with the polarization of the lenses also at right angles, insuring that each eye saw only its own image. Such Vectographs, unlike the red-and-green anaglyphs, could be in full color, which gave them a special appeal.

Then there were lenticular stereograms, in which the two images were printed in alternating narrow vertical bands covered by clear, ridged plastic. The ridges served to transmit each set of images to the proper eye, elimi-

---

I spoke to him about seeing faces as hollow masks, he was very surprised. With something as familiar and crucial as a face, he thought, probabilities and context would weigh the odds heavily against such a radical misperception. I agreed, but could not gainsay my own experience, and Gregory had to concede that such an improbable phenomenon might indeed occur in someone who is strongly biased towards binocular cues.

nating the need for any special glasses. I first saw a lenticular stereogram just after the war, in the London Tube—an advertisement, as it happened, for Maidenform bras. I wrote to Maidenform, asking if I could have one of their advertisements, but got no reply; they must have imagined I was a sex-obsessed teenager, rather than a simple stereophile.

Finally, in the early 1950s, there were 3-D films (like the Madame Tussauds horror film, **House of Wax**), which one would look at through red-and-green or Polaroid glasses. As cinema, some of these were awful—but a few, like **Inferno**, were very beautiful and used stereo photography in an exquisite, delicate, unintrusive way.

Over the years, I amassed a collection of stereograms and books about stereoscopy. I became an active member of the New York Stereoscopic Society, and at our meetings I encountered other stereo buffs. We stereo enthusiasts subscribe to stereo magazines, and some of us attend stereo conventions. The most ardent take their stereo cameras and go on "stereo weekends." Most people are not particularly conscious of what stereoscopy adds to their visual world, but we revel in it. While some may not notice any big difference if they close one eye, we stereophiles are sharply aware of a great change, as our world suddenly loses its spaciousness and depth and becomes as flat as a playing card. Perhaps our stereoscopy is more acute; perhaps we live, subjectively, in a deeper world; or perhaps we are simply more aware of it, as others may be more attuned to color or shape. We want to understand how stereoscopy works. The problem is not a trivial one, for if one can understand stereoscopy, one can understand not only a simple and brilliant visual stratagem but something of the nature of visual awareness, and of consciousness itself.

One has to lose the use of an eye for a substantial period to find how life is altered in its absence. Paul Romano, a sixty-eight-year-old retired pediatric ophthalmologist, recounted his own story in the **Binocular Vision & Strabismus Quarterly.** He had suffered a massive ocular hemorrhage, which caused him to lose nearly all sight in one eye. After a single day of monocular vision, he noted, "I see items but I often don't recognize them: I have lost my physical localization memory. . . . My office is a mess. . . . Now that I have been reduced to a two-dimensional world I don't know where anything is."

The next day he wrote, "Things are not the same at all monocularly as they were binocularly. . . . Cutting meat on the plate—it is difficult to see fat and gristle that you want to cut away. . . . I just don't recognize it as fat and gristle when it only has two dimensions."

After almost a month, though Dr. Romano was becoming less clumsy, he still had a sense of great loss:

> Although driving at normal speed replaces the loss of depth perception with motion stereopsis, I have lost my spatial orientation. There is no longer the feeling I used to have of knowing exactly where I am in space and the world. North was over here before—now I don't know where it is. . . . I am sure my dead reckoning is gone.

His conclusion, after thirty-five days, was that "even though I adapt better to monocularity every day, I can't see spending the rest of my life in this way. . . . Binocular stereoscopic depth perception is not just a visual phenomenon. It is a way of life. . . . Life in a two-dimensional world is very different from that in a three-dimensional world and very inferior." As the weeks passed, Dr. Romano became more at home in his monocular world, but it was with enormous relief that, after nine months, he finally recovered his stereo vision.

In the 1970s, I had my own experience with losing stereoscopy when I was put in a tiny windowless room in a London hospital, following surgery for a ruptured quadriceps tendon. The room was scarcely bigger than a prison cell, and visitors complained of it, but I soon accommodated and even enjoyed it. The effects of its limited horizon did not become apparent to me until later, as I described in **A Leg to Stand On:**

> I was moved into a new room, a new spacious room, after twenty days in my tiny cell. I was settling myself, with delight, when I suddenly noticed something most strange. Everything close to me had its proper solidity, spaciousness, depth—but everything farther away was totally flat. Beyond my open door was the door of the ward opposite; beyond this a patient seated in a wheelchair; beyond him, on the windowsill, a vase of flowers; and beyond this, over the road, the gabled windows of the house opposite—and all this, two hundred feet perhaps . . . seemed to lie like a giant Kodachrome in the air, exquisitely colored and detailed, but perfectly flat.
>
> I had never realized that stereoscopy and

spatial judgment could be so changed after a mere three weeks in a small space. My own stereoscopy had returned, jerkily, after about two hours, but I wondered what happened to prisoners, confined for much longer periods. I had heard stories of people living in rain forests so dense that their far point was only six or seven feet away. If they were taken out of the forest, it was said, they might have so little idea or perception of space and distance beyond a few feet that they would try to touch distant mountaintops with their outstretched hands.[5]

---

5. In **The Forest People,** Colin Turnbull described driving with a Pygmy man who had never left the jungle before:

> He saw the buffalo, still grazing lazily several miles away, far down below. He turned to me and said, "What insects are those?" At first I hardly understood; then I realized that in the forest the range of vision is so limited that there is no great need to make an automatic allowance for distance when judging size. . . . When I told Kenge that the insects were buffalo, he roared with laughter and told me not to tell such stupid lies. . . . As we got closer, the "insects" must have seemed to get bigger and bigger. Kenge kept his face glued to the window, which nothing would make him lower. I was never able to discover what he thought was happening—whether he thought that the insects were changing into buffalo, or that they were miniature buffalo growing rapidly as we approached. His only comment was that they were not real buffalo, and he was not going to get out of the car again until we left the park.

· · ·

When I was a neurology resident in the early 1960s, I read the remarkable papers of David Hubel and Torsten Wiesel on the neural mechanisms of vision. Their work, which later won a Nobel Prize, revolutionized our understanding of how mammals learn to see, in particular of how early visual experience is critical for the development of special cells or mechanisms in the brain needed for normal vision. Among these are the binocular cells in the visual cortex, which are necessary to construct a sense of depth from retinal disparities. Hubel and Wiesel showed, in animals, that if normal binocular vision was rendered impossible by a congenital condition (as in Siamese cats, which are often born cross-eyed) or by experiment (cutting one of the muscles to the eyeballs, so that the subjects became walleyed), these binocular cells would fail to develop and the animals would permanently lack stereoscopy. A significant number of people develop similar conditions—collectively known as strabismus, or squint—a misalignment sometimes too subtle to attract notice but sufficient to interfere with the development of stereo vision.

Perhaps 5 or 10 percent of the population, for one reason or another, have little or no stereo vision, though they are often not aware of this and may learn it only after careful examination by an ophthalmologist or optometrist.[6] Yet there are many accounts of stereo-blind people who nonetheless achieve remarkable feats of visuo-motor coordination. Wiley Post, the first person to fly solo around the world, as famous in the 1930s as Charles Lindbergh, did so after losing an eye in his mid-twenties. (He went on to become a pioneer of high-altitude flight and invented a pressurized flight suit.) A number of professional athletes have been blind in one eye, and so was at least one eminent ophthalmic surgeon.

Not all stereo-blind people are pilots or world-class athletes, and some may have difficulty judging depth, threading needles, or driving—but by and large they manage to get along pretty well using only monocular cues.[7]

---

6. More rarely, stereopsis may be lost, sometimes suddenly, with a stroke or other damage to the visual cortex. Macdonald Critchley, in his book **The Parietal Lobes,** also refers to the opposite condition as a rare consequence of cerebral lesions in the early visual cortex: an enhancement of stereo vision "whereby near objects seem to be abnormally close, and distant objects seem to be much too far away." Enhancement or loss of stereo vision can also occur transiently in a migraine aura or with certain drugs.

7. A number of people with misaligned eyes may not only lack stereo vision but have double vision or shimmering effects,

And those who have never had stereopsis but manage well without it may be hard put to understand why anyone should pay much attention to it. Errol Morris, the filmmaker, was born with strabismus and subsequently lost almost all the vision in one eye, but feels he gets along perfectly well. "I see things in 3-D," he said. "I move my head when I need to—parallax is enough. I don't see the world as a plane." He joked that he considered stereopsis no more than a "gimmick" and found my interest in it "bizarre."[8]

I tried to argue with him, to expatiate on the

---

which can cause them problems with daily activities generally, and especially with reading or driving.

8. Photographers and cinematographers, concerned to create an illusion of three-dimensionality on a flat plane, must deliberately renounce their binocularity and stereoscopy, confining themselves to a one-eye, one-lens view, to better frame and compose their pictures.

In a 2004 letter to the editor of the **New England Journal of Medicine**, Harvard neurobiologists Margaret Livingstone and Bevil Conway suggested, after an examination of Rembrandt's self-portraits, that the painter was so walleyed as to be stereoblind, and that "stereoblindness might not be a handicap—and might even be an asset—for some artists." Subsequently they proposed, after looking at photographs of other artists, that many of them—de Kooning, Johns, Stella, Picasso, Calder, Chagall, Hopper, and Homer, among others—also seemed to have significant misalignment of the eyes and were perhaps also stereo-blind.

special character and beauty of stereopsis. But one cannot convey to the stereo-blind what stereopsis is like; the subjective quality, the quale, of stereopsis is unique and no less remarkable than that of color. However brilliantly a person with monocular vision may function, he or she is, in this one sense, totally lacking.

And stereopsis, as a biological strategy, is crucial to a diverse array of animals. Predators, in general, have forward-facing eyes, with much overlap of the two visual fields; prey animals, by contrast, tend to have eyes at the sides of their heads, which gives them panoramic vision, helping them spot danger even if it comes from behind. The hammerhead shark is a fearsome predator, partly because its bizarre head shape allows its forward-facing eyes a greater separation—a hammerhead is a living hyperstereoscope. Another astonishing strategy is found in the cuttlefish, whose wide-set eyes normally permit a large degree of panoramic vision but can be rotated forward by a special muscular mechanism when the animal is about to attack, giving it the binocular vision it needs for shooting out its tentacles with deadly aim.[9]

---

9. Walleyed people enjoy an unusually wide field of vision due to the divergence of their eyes and may hesitate to sacrifice this for an operation that might align their eyes cosmetically but

In primates like ourselves, forward-facing eyes have other functions. The huge, close-set eyes of lemurs serve to clarify the complexity of dark, dense foliage, which, if the head is kept still, is almost impossible to sort out without stereoscopic vision—and in a jungle full of illusion and deceit, stereopsis is indispensable in breaking camouflage. On the more exuberant side, aerial acrobats like gibbons might find it very difficult to leap from branch to branch without the special powers conferred by stereoscopy. A one-eyed gibbon might not fare too well—and the same might be true of a one-eyed shark or cuttlefish.

Stereoscopy is highly advantageous to such animals, despite its costs: the sacrifice of panoramic vision, the need for special neural and muscular mechanisms for coordinating and aligning the eyes, and, not least, the development of special brain mechanisms to compute depth from the disparities of the two visual images. Thus, in nature, stereoscopy is anything but a gimmick, even if some human beings manage, and may even enjoy certain advantages, without it.

---

fail to give them stereoscopy. Intriguingly, several such people have written to me that they are able to converge their eyes and achieve stereo vision briefly.

. . .

In December of 2004, I received an unexpected letter from a woman named Sue Barry. She reminded me how we had met, in 1996, at a shuttle-launch party in Cape Canaveral (her husband, Dan, was an astronaut). We had been talking about different ways of experiencing the world—how, for example, Dan and other astronauts would lose their orientation, their sense of "up" and "down," in the microgravity conditions of outer space and had to find ways of adapting. Sue then told me of her own visual world: since she had grown up cross-eyed, her eyes did not work in tandem, and so she viewed the world with one eye at a time, her eyes rapidly and unconsciously alternating. I asked if this was any disadvantage to her. No, she said, she got along perfectly well—she drove a car, she could play softball, she could do whatever anyone else could. She might not be able to see depth directly, as other people could, but she could judge it as well as anybody, using other cues.

I asked Sue if she could **imagine** what the world would look like if viewed stereoscopically. Sue said yes, she thought she could—after all, she was a professor of neurobiology, and she had read Hubel and Wiesel's papers and much else on visual processing, binocular vision, and stereopsis. She felt that this knowledge had given her a special insight into what she was missing—she knew what stereopsis must be like, even if she had never experienced it.

But now, nearly nine years after our initial conversation, she felt compelled to write to me about this question:

> You asked me if I could imagine what the world would look like when viewed with two eyes. I told you that I thought I could. . . . But I was wrong.

She could say this because now she **had** stereopsis—and it was beyond anything she could have imagined. She went on to give me details of her visual history, starting with her parents noticing that she was cross-eyed a few months after she was born:

> The doctors told them that I would probably outgrow the condition. This may have been the best advice at the time. The year was 1954, eleven years before David Hubel and Torsten Wiesel published their pivotal papers on visual development, critical periods, and cross-eyed kittens. Today, a surgeon would realign the eyes of a cross-eyed child during the "critical period" . . . in order to preserve

> binocular vision and stereopsis. Binocular vision depends on good alignment between the two eyes. The general dogma states that the eyes must be realigned in the first year or two. If surgery is performed later than that, the brain will have already rewired itself in a way that prevents binocular vision.

Sue did have operations to correct her strabismus, first on the muscles of the right eye, when she was two, and then of the left eye, and finally of both eyes, when she was seven. When she was nine, her surgeon told her that she could now "do anything a person with normal vision could do except fly an airplane." (Wiley Post, apparently, had already been forgotten by the 1960s.) She no longer looked cross-eyed to a casual observer, but she was half aware that her eyes were still not working together, that there was still something amiss, though she could not specify what it was. "No one mentioned to me that I lacked binocular vision, and I re-mained happily ignorant of the fact until I was a junior in college," she wrote. Then she took a course in neurophysiology:

> The professor described the development of the visual cortex, ocular dominance columns, monocular and binocular vision, and experiments done on kittens reared with artificial

> strabismus. He mentioned that these cats probably lacked binocular vision and stereopsis. I was completely floored. I had no idea that there was a way of seeing the world that I lacked.

After her initial astonishment, Sue began to investigate her own stereo vision:

> I went to the library and struggled through the scientific papers. I tried every stereo vision test that I could find and flunked them all. I even learned that one was supposed to see a three-dimensional image through the View-Master, the toy stereo viewer that I had been given after my third operation. I found the old toy in my parents' home, but could not see a three-dimensional image with it. Everyone else who tried the toy could.

At this point, Sue wondered whether there might be any therapy by which she could acquire binocular vision, but "the doctors told me that it would be a waste of my time and money to attempt vision therapy. It was simply too late. I could only have developed binocular vision if my eyes had been properly aligned by age two. Since I had read Hubel and Wiesel's work on visual development and early critical periods, I accepted their advice."

. . .

Twenty-five years passed—years in which Sue married and raised a family while pursuing an academic career in neurobiology. Though she had some difficulties with driving—merging on entrance ramps to freeways, she found it hard to estimate the speed of oncoming cars—she got along generally quite well with her monocular ways of judging space and distance. Once in a while, she even teased binocular people:

> I took some tennis lessons with an accomplished pro. One day, I asked him to wear an eye patch so that he had to hit the ball using only one eye. I hit a ball to him high in the air and watched this superb athlete miss the ball entirely. Frustrated, he ripped off the eye patch and threw it away. I am ashamed to admit it, but I enjoyed watching him flounder, a sort of revenge against all two-eyed athletes.

But when Sue was in her late forties new problems began:

> It became increasingly difficult to see things at a distance. Not only did my eye muscles fatigue more quickly, but the world appeared

> to shimmer when I looked in the distance. It was hard to focus on the letters on street signs or distinguish whether a person was walking toward or away from me. . . . At the same time, my glasses, used for distance vision, made me far-sighted. In the classroom, I could not read my lecture notes and see the students at the same time. . . . I decided it was time to get bifocals or progressive lenses. I was determined to find an eye doctor who would give me both progressive lenses to improve my visual acuity and eye exercises to strengthen my eye muscles.

She consulted Dr. Theresa Ruggiero, a developmental optometrist, who found that Sue's eyes were developing various forms of imbalance—this sometimes happens after surgery for strabismus—so that the reasonable vision she had enjoyed for decades was now being undermined.

> Dr. Ruggiero confirmed that I saw the world monocularly. I only used two eyes together when looking within two inches of my face. She told me that I consistently misjudged the location of objects when viewing them solely with my left eye. Most importantly, she discovered that my two eyes were misaligned vertically. The visual field of my left eye was

> about three degrees above that of my right. Dr. Ruggiero placed a prism in front of my right lens that shifted the entire visual field of the right eye upward. . . . Without the prism, I had trouble reading the eye chart on a computer screen across the room because the letters appeared to shimmer. With the prism, the shimmer was greatly reduced.

("Shimmer," Sue later explained, was perhaps too mild a term, for it was not like the shimmer one might see with a heat haze on a summer day—it was, rather, a rapid, dizzying oscillation of several times a second.)

Sue got her new eyeglasses, complete with the prism, on February 12, 2002. Two days later, she had her first vision-therapy session with Dr. Ruggiero—a long session in which, using Polaroid glasses to allow a different image to be presented to each eye, she attempted to fuse the two pictures. At first, she did not understand what "fusion" meant, how it was possible to bring the two images together; but after trying for several minutes she found she was able to do it, though only for a second at a time. Although she was looking at a pair of stereo images, she had no perception of depth; nevertheless, she had made the first step, achieving "flat fusion," as Dr. Ruggiero called it.

Sue wondered whether, if she could hold her eyes aligned for longer, this would allow not just flat fusion but stereo fusion, too. Dr. Ruggiero gave her further exercises to stabilize her tracking and hold her gaze, and she worked on these exercises diligently at home. Three days later, something odd occurred:

> I noticed today that the light fixture that hangs down from our kitchen ceiling looks different. It seems to occupy some space between myself and the ceiling. The edges are also more rounded. It's a subtle effect but noticeable.

In her second session with Dr. Ruggiero, on February 21, Sue repeated the Polaroid exercise and tried a new one, using colored beads at different distances on a string. This exercise, known as the Brock string, taught Sue to fixate both eyes on the same point in space, so that her visual system would not suppress the images from one eye or the other but would fuse them together. The effect of this session was immediate:

> I went back to my car and happened to glance at the steering wheel. It had "popped out" from the dashboard. I closed one eye, then the other, then looked with both eyes again,

> and the steering wheel looked different. I decided that the light from the setting sun was playing tricks on me and drove home. But the next day I got up, did the eye exercises, and got into the car to drive to work. When I looked at the rear-view mirror, it had popped out from the windshield.

Her new vision was "absolutely delightful," Sue wrote. "I had no idea what I had been missing." As she put it, "Ordinary things looked extraordinary. Light fixtures floated and water faucets stuck way out into space." But it was "also a bit confusing. I don't know how far one object should 'pop out' in front of another for a given distance between the two objects. . . . [It is] a bit like I am in a fun house or high on drugs. I keep staring at things. . . . The world really does look different." She included some excerpts from her diary:

> February 22: I noticed the edge of the open door to my office seemed to stick out toward me. Now, I always knew that the door was sticking out toward me when it was open because of the shape of the door, perspective and other monocular cues, but I had never seen it in depth. It made me do a double take and look at it with one eye and then the other in order to convince myself

that it looked different. It was definitely out there.

When I was eating lunch, I looked down at my fork over the bowl of rice and the fork was poised in the air in front of the bowl. There was space between the fork and the bowl. I had never seen that before. . . . I kept looking at a grape poised at the edge of my fork. I could see it in depth.

March 1: Today, I was walking by the complete horse skeleton in the basement of the building where I work, when I saw the horse's skull sticking out so much, that I actually jumped back and cried out.

March 4: While I was running this morning with the dog, I noticed that the bushes looked different. Every leaf seemed to stand out in its own little 3-D space. The leaves didn't just overlap with each other as I used to see them. I could see the SPACE between the leaves. The same is true for twigs on trees, pebbles on the road, stones in a stone wall. Everything has more texture.

Sue's letter continued in this lyrical vein, describing experiences utterly novel for her, beyond anything she could have imagined or inferred before. She had discovered for herself that there is no substitute for experience, that there is an unbridgeable gulf between what Bertrand Russell called "knowledge by description" and "knowledge by acquaintance," and no way of going from one to the other.

One would think that the sudden appearance of an entirely new quality of sensation or perception might be confusing or frightening, but Sue seemed to adapt to her new world with remarkable ease. She was startled and disoriented at first, and had to calibrate her new visual perception of depth and distance with her actions and movements. But for the most part she felt entirely and increasingly at home with stereoscopy. Though she continues to be conscious of the novelty of stereo vision and indeed rejoices in it, she also feels now that it is "natural"—that she is seeing the world as it really is, as it should be. Flowers, she says, seem "intensely real, inflated," where they were "flat" or "deflated" before.

Sue's acquisition of stereoscopy after almost half a century of being stereo-blind has also been of great practical benefit. Driving is easier; threading a needle, too. When she looks down into her binocular microscope at work, she can see paramecia swimming at different levels, and see this directly, rather than inferring it by refocusing the microscope up or down. And it is a continuing source of enthrallment:

> At seminars . . . my attention is completely captivated by the way an empty chair displays itself in space, and a whole row of empty chairs occupies my attention for minutes. I would like to take a whole day just to walk around and LOOK. I did escape today for an hour to the college greenhouse just to look at the plants and flowers from all angles.

Most of the phone calls and letters I receive are about mishaps, problems, losses of various sorts. Sue's letter, though, was a story not of loss and lamentation but of the sudden gaining of a new sense and sensibility and, with this, a sense of delight and jubilation. Yet her letter also sounded a note of bewilderment and reservation: she did not know of any experience or story like her own and was perplexed to find, in all she had read, that the achievement of stereoscopy in adult life was "impossible." Had she always had binocular cells in her visual cortex, she wondered, just waiting for the right input? Was it possible that the critical period in early life was less critical than generally thought? What did I make of all this?

I mulled over Sue's letter for a few days and discussed it with several colleagues, including Bob Wasserman, an ophthalmologist, and Ralph Siegel, a vision physiologist.[10] A few weeks later, in February of 2005, the three of us went to see Sue at her home in Massachusetts, bringing along ophthalmological equipment and various stereoscopes and stereograms.

Sue welcomed us and, as we chatted, showed us some childhood photos, since we were interested in trying to reconstruct her early visual history. Her childhood strabismus, prior to surgery, was quite clear in the photographs. Had she ever been able to see in three dimensions? we asked. Sue thought for a moment and answered yes, perhaps—very occasionally, as a child, lying in the grass, she might suddenly see, for a second or two, a blade of grass stand out from its background; she had almost forgotten about this until we quizzed her. The grass would have to be very close to her eyes, within inches, requiring her (like any of us) to cross her eyes. So there was a suggestion that the potential for stereopsis was there and could

10. Together, the three of us had collaborated on several cases, including that of the "colorblind painter," who suddenly lost all ability to see in color, and that of Virgil, a man blind nearly from birth whose sight had been restored after almost fifty years of blindness. (Both of these case histories, "The Case of the Colorblind Painter" and "To See and Not See," were published in **An Anthropologist on Mars.**)

be brought out if she moved her eyes into the proper position for stereo viewing.

Sue had written in her letter, "I think, all my life, I have desired to see things in greater depth, even before I knew I had poor depth perception." This strange, poignant remark made me wonder whether she had retained some dim, barely conscious memory of having once seen things in greater depth (for she would have no sense of loss or nostalgia for something she had never had). It was important to test her with special stereograms that had no cues or clues as to depth—no perspective or occlusion, for example. I had brought one stereogram with lines of print—unrelated words and short phrases—that, if viewed stereoscopically, appeared to be on seven different planes of depth but, if viewed with one eye or without true stereo vision, appeared to be on the same plane. Sue looked at this picture through the stereoscope and saw it as a flat plane. It was only when I prompted her by telling her that some of the print was at different levels that she looked again and said, "Oh, now I see." After this, she was able to distinguish all seven levels and put them in the correct order.

Given enough time, Sue might have been able to see all seven levels on her own, but such "top-down" factors—knowing or having an idea of what one should see—are crucial in many aspects of perception. A special attention, a special searching, may be necessary to re-inforce a relatively weak physiological faculty. It seems likely that such factors are strongly operative with Sue, especially in this type of test situation. Her difficulties in real life are much less, because every other factor here—knowledge, context, and expectation no less than perspective, occlusion, and motion parallax—helps her experience the three-dimensional reality around her.

Sue was able to see depth in the red-and-green drawings I had brought. One of these images—an impossible three-pronged tuning fork such as M. C. Escher might have drawn, with three tines of increasing heights—Sue found "spectacular"; she saw the top of the uppermost prong as three or four centimeters above the plane of the paper. Yet Sue had spoken of herself as having only a "shallow" stereoscopy, and indeed, Bob and Ralph both saw the uppermost prong as about twelve centimeters above the plane of the paper, while I saw it as five centimeters higher still.

I found this surprising, because we were all the same distance from the drawing, and I had imagined that there would be, by a sort of neural trigonometry, a fixed relationship between the disparity of the images and their perceived depth. Puzzled by this, I wrote to Shinsuke Shimojo, at Caltech, an expert on many aspects of visual perception. He brought out, in his reply, that when one looks at a stereogram, the computational process in the brain is based not solely on the binocular cue of disparity but also on monocular cues such as size, occlusion, and motion parallax. The monocular cues may work against the binocular ones, and the brain must balance one set of cues against the other to arrive at a weighted average. This final result will be different in different individuals, because there is huge variation, even in the normal population: some people rely predominantly on binocular cues, others on monocular cues, and most use some combination of both. In looking at a stereo picture such as the tuning fork, a strongly binocular person will see unusual stereo depth; a monocularly oriented person will see much less depth; and others, relying on both binocular and monocular cues, will see something in between. Shimojo's formulation gave substance to the obstinate belief held by many of us in the New York Stereoscopic Society that we lived in a "deeper" world, visually, than the majority of people.[11]

---

11. If a stereo photograph is flashed on a screen for as little as twenty milliseconds, a person with normal stereoscopy can perceive some stereo depth straightaway. But what one sees in a

. . .

Later in the day, we paid a visit to Sue's optometrist, Dr. Theresa Ruggiero, who described how Sue had first consulted her, in 2001. Sue had complained then of eyestrain, especially when driving, impaired clarity, and a disconcerting jumping or flickering of images—but had not mentioned her lack of stereoscopy.

Dr. Ruggiero herself was greatly pleased, she said, when, immediately after achieving flat fusion, Sue experienced stereoscopy. The conscious effort and act of moving her eyes into position for binocular fusion, Ruggiero speculated, may have been critical for Sue's breakthrough. And she stressed, over and above the initial achievement of stereoscopy, Sue's ad-

---

flash is not the full depth; the perception of this requires several seconds, even minutes, in which the picture seems to deepen as one continues to gaze at it—it is as if the stereo system takes a certain time to warm up, to come to its full capacity. Such a deepening seems peculiar to the stereo system (colors, by contrast, do not normally become more pronounced as one looks at them). The underlying cause for this is unknown, though it has been suggested that it entails the recruitment of additional binocular cells in the visual cortex.

(There is, additionally, a clear practice effect, so that people who exercise their stereo powers—for example, by working with a binocular microscope—may experience striking improvements in stereo acuity and stereo depth over a longer period. Here, too, the underlying mechanism is unknown.)

venturous and positive reaction to this and her fierce determination to hold on to it and enhance it, however much work it might entail.

And it did indeed entail, and still entails, a great deal of work—taxing fusion exercises for at least twenty minutes every day. With these exercises, Sue found that she was starting to perceive depth at greater and greater distances, where at first she had seen depth only close up, as with the steering wheel. She continued to have jumps of improvement in her stereo acuity, so that she was able to see depth with smaller and smaller disparities—but when she stopped the therapy for six months, she quickly regressed. This upset her deeply, and she resumed the eye exercises, working on them every day, "religiously."

Sue uses a kinetic metaphor for her learning to use stereo vision, comparing it to learning to walk again. "I had to develop a new choreography for my own eye movements," she recently wrote, "how to move my eyes in harmony, before I could tap into latent binocular circuits and see in stereo depth."

Sue has continued to work very hard on her stereo perception and stereo acuity, and her perception of stereo depth is again on the increase. Moreover, she has developed a skill she did not have when we initially visited her: the ability to see random-dot stereograms. At first glance, these do not appear to contain any images at all. But as one continues to gaze at them through the stereoscope, one becomes aware of a strange sort of turbulence among the dots, and then a startling illusion—an image, a shape, whatever—suddenly appears far above or far below the plane of the paper. This illusion takes some practice, and many people, even those with normal binocular vision, are not able to get it. But it is the purest test of stereoscopic vision, for there are no monocular cues whatever; only by stereoscopically fusing thousands of seemingly random points as seen by two eyes can the brain construct a three-dimensional image.[12]

David Brewster, a nineteenth-century scientist who was inspired by Wheatstone's work, observed a related form of stereo illusion. Gazing at wallpaper with small repetitive motifs, he observed that sometimes, with the proper convergence or divergence of gaze, the patterns might quiver or shift and then jump into star-

---

12. Bela Julesz, the remarkable researcher who studied random-dot stereoscopy, spoke of "cyclopean vision," and regarded it as entailing neural mechanisms over and above those employed in ordinary stereo vision. This too is suggested by the fact that it may take a minute or more to "get" random-dot stereograms, where ordinary stereograms can be seen instantly.

tling stereoscopic relief, seeming to float in front of or behind the wallpaper.[13] Brewster wrote about these stereoscopic illusions, and believed he was the first to observe them—although it seems likely that such "autostereograms" have been experienced for millennia, with the repetitive patterns of Islamic art, Celtic art, and the art of many other cultures. Medieval manuscripts such as the Book of Kells or the Lindisfarne Gospels, for example, contain exquisitely intricate designs done so exactly that whole pages can be seen, with the unaided eye, in stereoscopic relief. (John Cisne, a paleobiologist at Cornell, has suggested that such stereograms may have been "something of a trade secret among the educated élite of the seventh- and eighth-century British Isles.")

In the past decade or two, elaborate autostereograms have been widely popularized in Magic Eye books. The illusions are single

13. Brewster also invented, around 1844, a simple handheld stereoscope using lenses (Wheatstone's mirror stereoscope was large and heavy and had to sit on a table). While Brewster was at first full of admiration for Wheatstone, he subsequently became jealous of his younger colleague and began publishing vindictive articles about him, pseudonymously. Finally, in 1856, in his otherwise charming book **The Stereoscope: Its History, Theory and Construction,** he attacked Wheatstone openly and denied him any claim to priority in the realm of stereoscopy.

images that one views without a stereoscope; but they contain horizontal rows of repeating "wallpaper" patterns that are slightly different. At first glance, all the patterns seem to be on the same level, but if one learns how to diverge or converge the eyes, letting each eye focus on a different row, then striking stereoscopic illusions appear. Sue loves these, and they have added another dimension to her newfound life in stereo: "I find these wallpaper autostereograms easy (and quite thrilling)," she recently wrote, "probably because I practice convergent and divergent fusion regularly."

In the summer of 2005, Bob Wasserman and I paid Sue another visit, this time in Woods Hole, Massachusetts, where she was running a fellowship program in neurobiology. She had mentioned to me that the bay there was sometimes full of luminous organisms, mostly tiny dinoflagellates, and that she enjoyed swimming among them. When we arrived, in the middle of August, we found that our timing was perfect; the water was aflame with the luminous creatures ("**Noctiluca scintillans**—I love the name," said Sue). After dark, we went down to the beach, armed with masks and snorkels. We could see the water sparkling from the shore, as if fireflies were in it, and when we immersed ourselves and moved our arms and legs in the water, clouds of miniature fireworks lit up around our limbs. When we swam, the night lights rushed past our eyes like the stars streaking past the **Enterprise** as it reaches warp speed. In one area, where the noctiluca were particularly dense, Bob said, "It's like swimming into a galaxy, a globular cluster."

Sue, overhearing this, said, "Now I see them in 3-D—they all seemed to twinkle in a flat plane before." Here there were no contours, no boundaries, no large objects to occlude or give perspective. There was no context whatsoever—it was like being immersed in a giant random-dot stereogram—and yet Sue now saw the noctiluca at different depths and distances, in three-dimensional space. We wanted to quiz her in more detail about the experience, but Sue, normally eager to talk about stereo vision, was mesmerized by the beauty of the scintillating organisms. "Enough thinking!" she said. "Give yourself to the noctiluca."

Struggling to find an analogy for her experience, Sue had suggested, in her original letter to me, that her experience might be akin to that of someone born totally colorblind, able to see only in shades of gray, who is suddenly given the ability to see in full color. Such a person, she wrote, "would probably be overwhelmed by the beauty of the world. Could they stop looking?" While I liked the poetry of Sue's analogy, I was unsure about the thought. (My friend and colleague Knut Nordby, who was completely colorblind, thought that to be given color as an "add-on" after a lifetime without it would be grossly confusing, and impossible to integrate with his already complete visual world. Color, he felt, would be unintelligible and have no associations, no meaning, for someone like him.)

Sue's experience of stereoscopy, however, was clearly not a gratuitous or meaningless addition to her visual world. After a brief confusion, she embraced the new experience and felt it not as an arbitrary add-on but as an enrichment, a natural and delicious deepening of her existing vision. But terms like "enrichment" or "deepening," Sue felt, did not begin to do justice to her acquisition of stereoscopy. It was not just a quantitative increase; it was something entirely novel. Stereoscopy, she maintains, is subjectively **different.**[14] This difference even extends

14. This view, which I share, seems to be in contradiction with the views of the great visual pioneer J. J. Gibson. In his 1950 book **The Perception of the Visual World**, he wrote, "If the gradient theory is correct, binocular vision simply takes its place as a determinant, but only one determinant, of visual space." Several eminent contemporary vision researchers hold

to the perception of two-dimensional representations such as photographs, movies, or paintings. Sue now finds these far more "realistic"; her now-activated stereo systems allow her to **imagine** space in a way she could not before.

David Hubel has followed Sue's case with interest and has corresponded with her and with me about it. He has pointed out that we are still quite ignorant of the cellular basis of stereoscopy. We do not know whether, even in animals, disparity-sensitive cells (the binocular cells specialized for stereoscopy) are present at birth (though Hubel suspects they are). We do not know what happens to these cells if there is strabismus and lack of binocular experience in early life or, most crucially, whether they can recover if people later learn to position their eyes for binocular fusion. With regard to Sue, he wrote, "It seems to me that [her regaining

---

similar views. Thus Dale Purves and R. Beau Lotto, in their book **Why We See What We Do,** write of "a seamless relationship" between the three-dimensional world we construct with one eye and its "augmentation" by stereopsis. Such views, while wholly consistent with a behavioral or empirical theory of vision, give no weight to the qualitative and subjective aspects of stereoscopy. Here one needs inside narratives, personal accounts of what it is like to suddenly gain stereo vision after a lifetime of stereo blindness (as Sue describes) or to suddenly lose it after a lifetime of seeing in stereo (as I describe in the following chapter).

of stereopsis] occurred too quickly for it to be due to a reestablishment of connections, and I rather would guess that the apparatus was there all along, and just required reestablishment of fusion to be brought out." But, he added, "that's just a guess!"

What emerges from Sue's experience is that there seems to be sufficient plasticity in the adult brain for these binocular cells and circuits, if some have survived the critical period, to be reactivated much later. In such a situation, though a person may have had little or no stereo vision that she can remember, the potential for stereopsis is nonetheless present and may spring to life—most unexpectedly—if good alignment of the eyes can be obtained. That this seems to have happened with Sue after a dormant period of almost fifty years is very striking.

Though Sue originally thought her own case unique, she has found, on the internet, a number of other people with strabismus and related problems who have unexpectedly achieved stereo vision through vision therapy. Their experiences, like Sue's, suggest that if one has even small islands of function in the visual cortex, there may be a fair chance of reactivating and expanding them in later life, despite a lapse of decades.

Whatever its neurological basis, the augmentation of Sue's visual world has effectively granted her an added sense, a circumstance that the rest of us can scarcely imagine. For her, stereopsis continues to have a quality of revelation. "After almost three years," she wrote, "my new vision continues to surprise and delight me. One winter day, I was racing from the classroom to the deli for a quick lunch. After taking only a few steps from the classroom building, I stopped short. The snow was falling lazily around me in large, wet flakes. I could see the space between each flake, and all the flakes together produced a beautiful three-dimensional dance. In the past, the snow would have appeared to fall in a flat sheet in one plane slightly in front of me. I would have felt like I was looking in on the snowfall. But now, I felt myself within the snowfall, among the snowflakes. Lunch forgotten, I watched the snow fall for several minutes, and, as I watched, I was overcome with a deep sense of joy. A snowfall can be quite beautiful—especially when you see it for the first time."

Postscript

Seven years after acquiring stereoscopy, Sue still delights in her "new" sense and finds her visual world infinitely richer for it. Since writing to me in 2004, she has continued to think about her own experiences and to reach out to many people in similar situations, as well as to vision researchers. In 2009, she published a beautiful and profound book about her experiences, **Fixing My Gaze: A Scientist's Journey into Seeing in Three Dimensions.**

Persistence of Vision

A Journal

ON DECEMBER 17, 2005, a Saturday, I had my usual morning swim and then decided to go to the movies. I arrived a few minutes early and took a seat in the back of the cinema—I had no intimation of anything unusual until the previews started. Then I immediately became conscious of a sort of fluttering, a visual instability, to my left. At first I thought it was the start of a visual migraine, but I soon realized that whatever it was affected only the right eye and must therefore be arising in the eye itself, and not in the visual cortex, the way a migraine would.

When the cinema screen went dark after the first preview, the spot that had been quivering to my left flared up like a white-hot coal, with spectral colors—turquoise, green, orange—at its edges. I was alarmed: was I having a hemorrhage into the eye, a blockage of the central retinal artery, a retinal detachment? I then became conscious of a blind spot within the incandescent area, for using just my right eye and looking to the left, where a line of little lights along the floor indicated a way out of the cinema, I found that all the forward ones were now "missing."

I felt panic rising. Would the dark area continue to enlarge until the right eye was completely blind? Should I leave at once? Go to an emergency room? Call my ophthalmologist friend, Bob? Or should I sit tight and see if the disturbance spontaneously resolved? The film started, but I paid little attention to it; I was entirely preoccupied with checking my vision every few seconds.

Finally, after about twenty minutes, I burst out of the cinema—perhaps everything would look fine once I got into daylight, the real world. But it didn't. The flaring had died down a little, but when I used only my right eye, a pie-shaped chunk of my visual field was still missing to the left. I walked, almost ran, back to my apartment and phoned Bob. He asked a few questions, suggested a couple of instant tests, then told me to get myself to an ophthalmologist immediately.

A couple of hours later I was in the ophthalmologist's consulting room. I told my story again, indicated the quadrant of blindness in my right eye. He listened carefully, looked noncommittal, and, after checking my visual fields, took his ophthalmoscope and peered into the eye. Then he put down the instrument, leaned back, and gazed at me, I thought, with different eyes. There had been a certain lightness or casualness in him before—we were not exactly friends, but we were colleagues, both medical men. Now, suddenly, I was in a quite different category. He spoke carefully, picking his words; his demeanor was one of seriousness and concern. "I see pigmentation," he said, "something behind the retina. It could be a hematoma, or it could be a tumor. If it's a tumor, it could be benign or malignant." He seemed to take a deep breath. "Let's look at the worst-case scenario," he continued. I cannot be sure what he said next, for a voice had started up in my head, shouting, "CANCER, CANCER, CANCER . . ." and I could no longer hear him. He said he would make arrangements for me to see Dr. David Abramson, a great expert on ocular tumors, as soon as possible.

Back in my apartment that evening, testing my right eye, I was startled to see that the horizontal bars on the air conditioner all seemed to be warped, converging and collapsing into one another, while the vertical bars diverged. I cannot now remember how I spent the rest of the weekend. I was very restless, I went for long walks, and when I was inside, I paced to and fro. The nights were especially bad—I had to knock myself out with sleeping pills.

DECEMBER 19, 2005: DIAGNOSIS

I was able to see Dr. Abramson first thing on Monday. Kate—she is my close friend, as well as my assistant—came with me, for moral support. Dr. Abramson was a quiet man, sober, measured, reserved, with a mischievous glint in his eye. "Nice to meet you," I said.

"We have met before," he answered, and reminded me that he had been one of my students, back in the 1960s. He had vivid memories of my teaching and some of my idiosyncrasies. He recalled that my class had been the only one in his medical school career to conclude each week with a general discussion over a cup of tea. How odd, I thought (as perhaps he did), that more than thirty-five years after being his mentor, I was now his patient.

He made a preliminary examination of my eyes, then put some drops in to dilate the pupils. This was followed by photography and an ultrasound examination of the retina. Little was said during these exams. Then we sat down in another, bigger room. Dr. Abramson brought out a large model of the eye, cut open to reveal its inner anatomy. Taking a hideous-looking black object—irregular, convoluted, like a little black cauliflower or cabbage—he placed it near the entry of the optic nerve. The meaning of this was clear: I had a tumor, a malignant one. I thought of how, in England, the judge dons a black cap before pronouncing a death sentence. The black cabbage had the same meaning. I felt that I had been given a death sentence.

"It's a melanoma," he confirmed, but immediately went on to say that ocular melanomas rarely metastasized—there was little chance of any spread beyond the eye. Nevertheless, one could not allow it to persist and grow, untreated, in the eye. Until fairly recently, the recommended procedure was removing the entire eye (he himself had done a thousand such enucleations over the years) but now, it was felt, radiation could be just as effective, allowing one to keep the eye and its remaining vision. Dr. Abramson had barely got this out before I asked how soon this radiation could be done: tomorrow? He said there would be a three-week delay—the Christmas and New

Year's holidays were coming up—but there would be no significant growth of the tumor in this time, he reassured me; these things tended to be very slow-growing. It would take some time to fashion the radioactive plaque itself, tailoring it to focus the radiation precisely on the tumor. Then the plaque would be attached to the side of my eye, which would require disconnecting one of the eye muscles. In a second operation a few days later, the plaque would be removed and the muscle reconnected.

My tumor must have taken some time to reach this size, he added—had I observed any defect in my visual field in the months before? Alas, I had never checked it. I had noticed nothing amiss until two days before, in the cinema, and then the odd visual distortions, the warping of horizontals and verticals, over the weekend. This was due to the swelling and distortion of the retina, Dr. Abramson said, and it would disappear as the tumor, and the edema associated with it, yielded to treatment. But if the distortions grew worse, he suggested, I might consider wearing an eye patch for a few weeks until they subsided.

Ocular melanomas were virtually all sensitive to radiation, he continued. There was a very good chance that the tumor could be killed by the radiation, followed, if need be, by lasering. Unfortunately, my tumor was in a bad location—scarcely more than a hundred cells, a single millimeter, from the fovea, the part of the retina that one fixates with, where visual acuity is greatest. But if the tumor could be stopped in its tracks, he said, I would retain, for a while, the 20/20 vision I had always enjoyed in this eye. Later there might be some loss of vision, due to the belated effects of the radiation. Still, I should have a substantial "window"—perhaps years—of good vision before this occurred.

I said to Dr. Abramson, "I guess you have to give news like this to many patients." I asked how I had seemed to take the news. Very calmly, he said, but it would need some digesting.

DECEMBER 19, 2005

I wake from a nightmare. The moment I open my right eye, I perceive something is wrong. The Darkness has inched forward—I can hardly see anything now to the left. I am calm and rational on the surface; I know that, with David Abramson, I am in the best possible hands, but I feel a terrified child, a child screaming for help, inside me.

DECEMBER 21, 2005

Having cancer, any cancer, means an instant change in status, in one's life. The diagnosis is a threshold beyond which lies a lifetime, however long, of tests, treatments, vigilance—and always, whether conscious or unconscious, a sense of reservation about the future. Today, the first day of winter, liver function tests are to be done. Has the beast spread to my liver? Does it have its claws in my vitals? Will I die of melanoma? The thought is in my mind all the while.

I have made a bargain with the tumor: you can have the eye, if you insist, so long as you leave the rest of me alone.

At Memorial Sloan-Kettering there is a special sidewalk marked "Reserved for Patients Going to MSK." I had occasionally noticed it when I visited people at the hospital. "Poor buggers," I would think as I saw people take it. Now it is the path I take myself.

Blood is drawn—will it be normal? Routine exam: pulse, blood pressure, etc. My blood pressure is up a bit, 150/80—it is normally under 120/70. The elevator to the X-ray suite seems to have a strange, trapezoidal shape, its walls converging inward to the back. Is this part of the funhouse world, the world of metrical and topological distortion, that I will have to traverse? Kate assures me that this time, at least, it is not my eyes. The elevator indeed has a trapezoidal shape.

After a round of tests and paperwork at the hospital, I go back to Dr. Abramson's office, a few blocks away. I am beginning to know the place and his staff, and they, now, are beginning to know me. I have joined a new club—the Ocular Melanoma Club of Greater New York (just as I belong to the Mineralogical Club of New York . . . and the New York Stereoscopic Society, of which I may soon become the only monocular member).

"December twenty-first, the first day of winter," I say to Kate.

"An auspicious day," she replies, trying to cheer me up. "The days start getting longer."

"Yours, perhaps," I remark darkly.

DECEMBER 22, 2005

4 A.M.: Woke. Cold. The fear. I open my right eye. The Darkness has grown again, is coming to encircle my little island of vision, my fixation point, my fovea. Soon it will be engulfed entirely.

10 A.M.: Vision much better. I think my

4 a.m. observation was related to the semi-darkness of my bedroom and the fact that (as I am learning) the blind area, the scotoma, varies with the illumination—it can get larger and even knock out central vision if the light is dim.

When I close my right eye, I again see brilliant lights, the blinding lights that herald blindness. A scalloped crescent, with a Technicolor edge, just above my fixation point.

DECEMBER 23, 2005

I find that if I use just my right eye, I cannot read—the lines are indistinct, slippery, grossly distorted; they waver from moment to moment. I had not realized that this would be upon me so soon. Perhaps I have avoided reading these last few days, or done it wholly with my left eye, without realizing it. I am tending to close my right eye when I read—this is unconscious, involuntary, almost automatic.

DECEMBER 24, 2005

Waking after a good night's sleep, and with the morning sun pouring in through my windows,

I forgot for a moment that I am now a "cancer victim." I felt well, and the visual symptoms were not intrusive. Feeling well is always a bit dangerous for me—it tempts me to excess. This morning at the pool, I swam for too long: an hour, mostly backstroke, but then several lengths of freestyle, which Dr. Abramson had advised against (as, perhaps, tending to cause the retinal edema to pool), followed by a half hour of vigorous exercises with mat and ball. It was at this point that my vision started to go again—testing my right eye an hour later, I found I could not read even the large headlines of the **New York Times.** This terrified me, showed me what loss of central vision was like.

Now, two and a half hours later, the edema is settling (if it **was** edema), though vision in the right eye is still swimmy: lines and surfaces snake and curve. I find it easier to put a patch on the right eye and use just the left, which at least has stable vision.

Inside the blazing, coruscating margin of the scotoma, involuntary imaging of all sorts—faces, figures, landscapes—is going on continually. I have had similar images briefly at the start of a migraine or before falling asleep, but never, as far as I can remember, continuous imagery as I have now.

DECEMBER 25, 2005

Everyone says "Merry Christmas!" and I reply in kind, but this is the darkest Christmas I have ever known. The **New York Times** today has pictures and stories of various figures who have died in 2005. Will I be among those figures in 2006?

Kate tries to remain upbeat. "Dr. Abramson said that this would not kill you," she said. "Whatever happens, we will deal with it." I am not so sure. The idea of blindness terrifies me, as does the thought that perhaps I will be among the unlucky one percent.

DECEMBER 30, 2005

8 A.M.: This morning when I opened my eyes, the dark cloud in the right eye was much larger. Sitting up and looking out the window with the right eye, I could hardly see the sky at all, and I found, looking up at the center of my ceiling fan, that three of its five blades were scarcely visible to my right eye—I could see just the stumps of the blades, close to my fixation point.

10 A.M.: Now, after being up for two hours, I find that the scotoma has retreated and that I

can see all but one blade. Position is important, since the edema seems to pool when I lie flat at night—perhaps I should sleep with my head propped up.

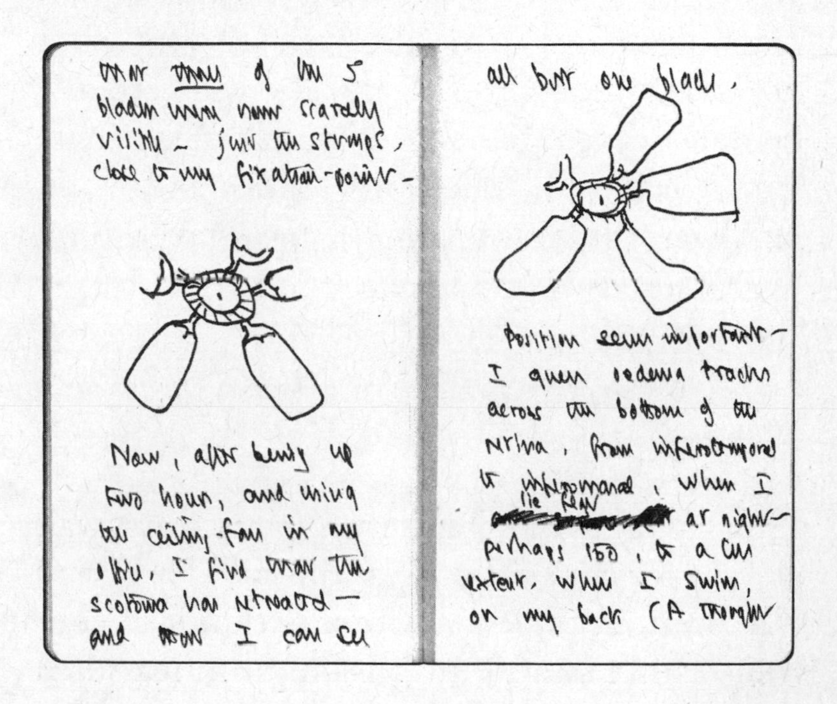

I find it difficult to concentrate, to compose myself. Difficult, too, to write—I have not written anything (other than brief letters) since completing a chapter on musicogenic epilepsy a week ago—though I have been thinking, at least, of synesthesia and music.

4 P.M.: Mood and energy much better! I have just written the greater part of "Colored Music," my chapter on synesthesia.

JANUARY 1, 2006

On this New Year's Day, I find myself entertaining fears and hopes, facing challenges of an entirely new kind. There is a small but significant chance that this will be my last year—but whether or not this is so, my life will certainly be transformed, has already been transformed, in a radical way. Questions of love and work, of what really matters most, have taken on a special intensity and urgency.

JANUARY 5, 2006

I am impatient and annoyed that I must wait so long for the surgery. Has this holiday period cost precious time, allowing the tumor to continue eating away at my vision? I am reassured that Dr. Abramson will do everything possible to kill this tumor, while preserving as much of my sight as possible. And I am glad to have met him again (though not in these circumstances). He is not only a brilliant physician but an extremely sensitive man—very important when dealing with people who have cancer. He never seems hurried or impatient. He listens carefully to what I say and responds with great delicacy and tact. I think he has my measure, as well as the melanoma's.

JANUARY 8, 2006

I slept fitfully last night, with dreams and anxieties about the eye, about vision—and, beyond this, about my life. Fears of every sort are rushing through my mind, mixed with (futile) regrets and recriminations that the tumor was not diagnosed earlier. Why did I not realize the import of those close-set wavy lines, the little stars and tussocks, which I had been seeing on the white ceiling of the swimming pool for the last few months whenever I did the backstroke? How could I be so absurd as to dismiss them as "fragments of migraine" or a reflection of my eyelashes in the goggles, when a moment's experiment would have shown me—as I found yesterday—that they were only to be seen with the right eye and equally visible without the goggles? I could, **should** have paid attention, questioned, sought clarification months ago.

Bob, however, feels that this would not have made an appreciable difference, but what is damnable—and here I am mad at my former eye doctor, at Kate, and at myself—is that my "annual" eye exam was somehow missed two years in succession, so I went thirty-two months without an eye exam. This delay could perhaps cost me my vision, even my life—but I must not think on this; must focus instead on how fortunate I am that the thing has been caught now and, as Dr. Abramson says, is wholly treatable.

JANUARY 9, 2006: SURGERY

10 A.M.: I am due to go to surgery in an hour or so; I do not know how conscious I will be, or want to be. With previous operations—shoulder and leg surgery—I was eager to know, almost to participate in the proceedings. This time I would like to be out, completely out. Kate and Bob are here with me and are trying to reassure and distract me.

5 P.M.: I **was**—happily, deliciously—out of it during the procedure. As the fentanyl took effect, the sciatica I have been plagued with for months disappeared, and I drifted into an unconsciousness deeper than the deepest sleep. When I came to, Dr. Abramson asked me a question or two to test my orientation and cognitive status. Where was I? What had been done? I replied that I was in the recovery room and that he had detached the lateral rectus muscle of the right eye and attached the plaque containing radioiodine (I-125, to be precise) to the sclera. I said that I was sorry it was not radioactive ruthenium instead of iodine (I have a thing for the platinum metals) but that 125, at least,

was memorable for being the smallest number that was the sum of two squares in two different ways. I startled myself as I said this; I had not thought it out before—it just jumped into my mind. (I realized, a few minutes later, that I was wrong—65 is the smallest such number.) I continued in a loquacious, slightly euphoric state and, for me, an unusually amiable and sociable one, chatting with all the nurses. Kate came in to visit me in the recovery room (she told me later that she had to reassure the nurses that my low pulse is normal, for I am a long-distance swimmer).

Now, six hours later, lying in bed, I see occasional sparkles, scintillations, in my right eye. I wonder if these are from particles or rays emitted by the radioiodine hitting my retina. (It makes me think back to the radioactive clock dials my Uncle Abe used to make, and how I would press these against my closed eyelids as a child and see similar scintillations . . . could this have played a part in causing my tumor?)

My eye is covered by a thick wad of gauze and a rigid eye patch to protect the eye from any jostling. There is a radioactivity warning sign on my door. People can only enter my room provided they obey instructions—and I cannot leave it. No children or pregnant women are permitted, and no one is allowed to kiss me for the days that my radioactive plaque is in place. I am not allowed to go home; I am under hospital arrest. I am "hot."

JANUARY 10, 2006

4 A.M.: Up, restless, can't sleep anymore. The patch presses on my eye, oppresses me (someone had the witty thought of bringing in a book called **The Blindfold**, by Siri Hustvedt), but my sciatica—which has tormented me for months—is still, mysteriously, in abeyance. The room is quiet, peaceful, undemanding, and I can gaze at the East River slowly moving by.

9 A.M.: Looking through the window, with my unpatched left eye, I am startled to see cars stuck in the branches of trees, like toys. With one eye occluded, I have no sensation of distance or depth whatever, a foretaste of what it will be like if I lose central vision in the right eye.

3 P.M.: Visitors and phone calls nonstop since this morning. Wonderful—but exhausting. Kate went out to find me some comfort food and came back with a bagel and whitefish; other friends have brought chocolates and fruit, matzoh ball soup, challah and schmaltz

herrings. It is herring and smoked fish I crave most when I am down. Between that and the hospital food, I am well stocked and quite happy to be alone now.

4 P.M.: A pall has descended over the city—a soft gray mist rendering the East River invisible and softening the blocky outlines of the buildings around me. A gentle, beautiful pall.

5 P.M.: A sudden stabbing pain in my eye, then a turmoil of raying purple forms, starfish, daisies, expanding outward from a multitude of separate points. This turmoil seems to fill the whole visual field. It fascinates and frightens me. Is something adrift, askew, amiss in the eye? Or is it my brain filling in, generating visions, in reaction to the cutting off of vision from the operated eye?

7 P.M.: Dr. Abramson came in for a long chat around six o'clock: How was I feeling generally? And what about the eye? I described my "visual storm," the starfish, etc. He thought it was probably a retinal reaction to the radiation. Picking up on this, I mentioned my thought—half serious, half joking—that the radioactivity in my eye might be strong enough to make my fluorescent minerals glow. Perhaps I could light them up by fixing my radioactive eye, my **rays** on them—it would be quite a nice party trick! Dr. Abramson was amused, said I should

ask Kate to bring the minerals in, and that he would take off the dressing so I could try.

He spoke, too, of how, in a few weeks, it might be a good idea to laser the retina and kill any malignant cells that might have survived the radiation. But my tumor is almost on top of the fovea, and if the fovea is destroyed, I will lose all central vision. He wondered about a compromise: lasering the two-thirds of the tumor farthest from the fovea, but keeping well clear of the fovea itself. He mentioned some newer treatments as well: injections of a substance into the eye that may prevent the growth of blood vessels within the tumor and thus starve it of blood; and a new anti-melanoma vaccine, still experimental. But for the moment, this is all in the future, hypothetical; he hopes radiation and lasering will do the trick.

In the meantime, I still have another thirty-six hours until Thursday afternoon, when I will go into surgery again to have the radioactive plaque removed.

JANUARY 11, 2006

My good friend Kevin came by at 6:15 this morning, a startling but very welcome apparition, with his huge, bushy eyebrows. He had been making early rounds on his patients and was still in his white coat. "Look!" he said, pointing to the window, and I looked—and saw a most delicate, rosy dawn transfusing the night sky and then a smoky, almost Krakatoa-like sunrise over the East River.

My scotoma itself is not so much like a blind spot as like a window, through which I see strange buildings, figures moving, little scenes playing themselves before me. At other times, I see writing, jumbled letters that I cannot read—hieroglyphs or runes—all over the scotoma. Once I saw an immense circular segment with numbers on it, like part of a clock or an Aztec calendar. I have no power to influence any of these visions; they proceed autonomously and have no connection that I can discern with what I am thinking or feeling. The sparkles, the visual storms, may come from my retina, but these visions, surely, must come from a higher level, must be constructed by my brain, calling, if indirectly, on its stock of images.

If I have been looking at something and then close my eyes, I continue to see it so clearly that I wonder whether I have actually closed my eyes. A startling example of this happened a few minutes ago when I was in the bathroom. I had washed my hands, was staring at the wash-basin, and then, for some reason, closed my left eye. I still saw the washbasin, large as life. I went back into my room, thinking that the dressing over the right eye must be absolutely transparent! This was my first thought and, as I realized a moment later, an absurd one. The dressing was anything **but** transparent—it was a great wodge of plastic, metal, and gauze half an inch thick. And my eye, beneath it, still had one of its muscles detached and was in no position to see anything. For the fifteen seconds or so that I had kept my good eye closed, I could not have been seeing anything at all. Yet I **did** see the washbasin—clear, bright, and real as could be. For some reason, the image on my retina, or in my brain, was not being erased in the normal way. And it was not a mere afterimage. Afterimages, for me at least, are brief and meager in the extreme—if I look at a lamp, I may see the glowing filament for a second or so—whereas this image was as detailed as reality itself. I continued to see the washbasin, the commode next to it, the mirror above it, the entire scene for a good fifteen seconds—a genuine persistence of vision. Something very odd was going on in my brain. I had never experienced such a phenomenon before. Was this—like my involuntary images, my hallucinations of patterns, of people—simply a consequence of being blindfolded in one eye? Or was it the angry, half-destroyed, cancered retina, now in a blaze of radiation from the radioiodine, sending strange wild signals to my brain?

JANUARY 12, 2006

8 A.M.: This afternoon, after precisely seventy-six hours, the radioactive implant will be removed, the disconnected eye muscle reconnected; and if all goes as it should, I will be released from the hospital tomorrow.

6 P.M.: I thought this surgery would be as sweet and painless as the first one, but when the anesthesia wore off I had the worst pain I have ever known—it made me gasp. I can avoid it only by keeping the eye completely still; the slightest movement seems to tear at the just reconnected eye muscle.

7 P.M.: Dr. Abramson came in to check my eye. He took off the patch; everything was very blurry, but this, he said, would clear in a day or so. He gave me careful instructions about putting drops in the eye several times a day, said I should not worry if I had transient double vision, and that I should feel free to call him, day or night, if I felt anything untoward was happening.

There is an unpleasant feeling of stickiness,

crustiness, in the eye, perhaps from all the eyedrops. I have to fight against the impulse to rub it.

MIDNIGHT: Finally, the pain is beginning to be tolerable. Over the past six hours I have had huge doses of Percocet and Dilaudid. Nothing seemed to touch the pain, until, an hour ago, Dr. Abramson ordered a whopping dose of Tylenol. Oddly, this did the trick, when the opiates had failed to help.

JANUARY 13, 2006

I came home this morning. Patients are usually glad to get out of the hospital, but I was rather sorry to leave. In the hospital, I was surrounded by attentive people, catering to every need; I was visited constantly, pampered. And now all this has gone, and I am back in my apartment, alone. I can't go out—there has been a heavy snowfall, the streets are icy—and I dare not go for a walk with, in effect, only one eye at the moment.

JANUARY 15, 2006

7 A.M.: There was a snowstorm, a howling gale, in the night, but it looks pretty, what I can see

of it now. Mornings are worst. I wake to a small, dim, bleary window of vision in my right eye, with streaks and blotches moving across it and gross distortion of horizontals and verticals, as one might get with a fish-eye lens.

10 A.M.: It has been almost a week since the surgery, and tired of staying inside, I ventured out, despite the snow, on a friend's arm. It is extremely cold, icy, windy outside. The wheels of cars spin helplessly, and we saw one car, parked on ice, actually blown forward an inch or two by a sudden, gale-force gust.

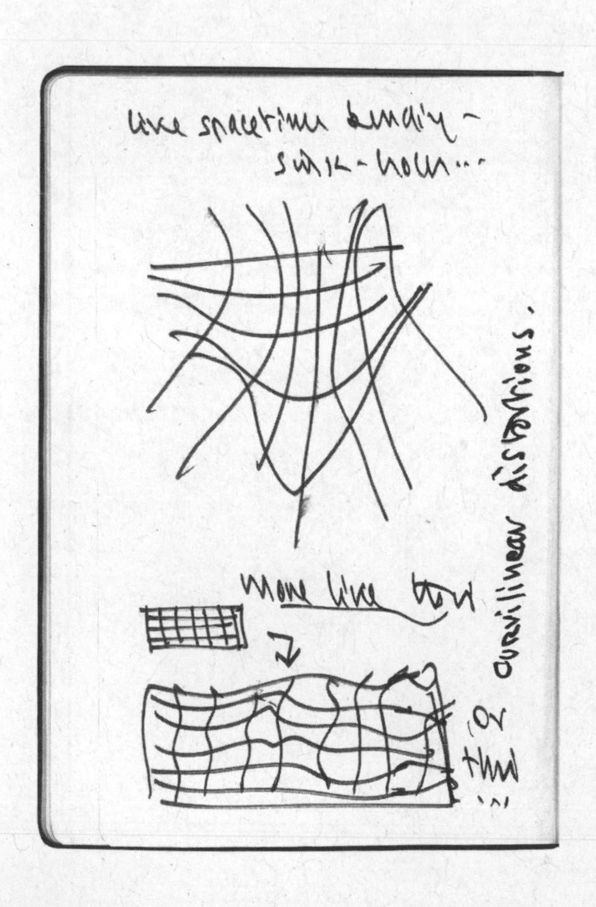

Everything in the right eye is swimmy, not only metaphorically but literally so—I am looking through a shifting film of fluid. The shapes of everything are fluid, moving, distorted. I imagine my retina almost afloat in the fluid pooling beneath it, changing shape like a jellyfish, or maybe a waterbed.

Looking through a window at a tall rectangular building across the street, I see it, as in a fun house, with its top or its middle (depending on where I fixate) splayed out and bulbous. This happens with all verticals; horizontals tend to

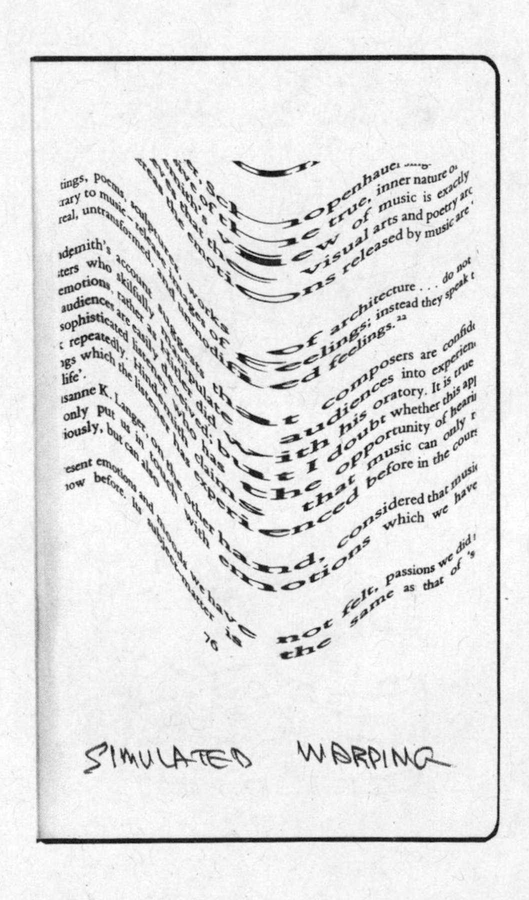

be squashed together. In the bathroom mirror, the upper part of my reflection is distorted—my head looks grotesquely flattened.

I am told that these effects come from edema beneath the retina and will resolve in a few days. I cannot always believe this; I feel that something approaching blindness in my right eye has descended on me far faster than I (or anyone else) would have predicted. Along with this is the suspicion that there was a fatal delay between diagnosis and treatment. That in those three weeks, additional and irreversible damage was done, as vision declined from a smallish blind spot to a virtual obliteration of the whole upper hemisphere of vision. I cannot help feeling that the melanoma should have been treated as an emergency and radiated without delay. I am sure I am being irrational, I hope I am wrong on these matters—but they form a nucleus of distrust and suspicion, which can get blown up into a tornado of paranoia.

JANUARY 16, 2006

Have just written to Simon Winchester, telling him how much pleasure I have got from listening to the audiotape of his book **Outposts.**

I live in a world of words, and I need to read; much of my life is reading. This is not easy now, with my right eye being "out" for the moment, and the left eye with long-standing problems of its own. I was punched in the left eye as a boy, which produced a cataract, and its vision has been below par ever since. This didn't matter when my dominant right eye had 20/20 vision, but now it does. My regular reading glasses are not strong enough for my left eye; I have to use a magnifying glass, which makes reading much slower and prevents me from scanning whole pages at a time.

Wandered out with Kate to the bookstore to get some large-print books—dismayed to find that almost all their large-print books are how-to books or romance novels. I could hardly find a single decent book in the entire large-print section. It is as though the visually impaired are also regarded as intellectually impaired. I feel like writing a furious op-ed about this for the **Times.** Audiobooks have a larger range, but I have been a reader all my life, and am not fond of being read to, on the whole. The Simon Winchester was a pleasant exception to the rule.

JANUARY 17, 2006

Dr. Abramson cautioned me that while the retina is still swimming in edema, I may see clearly one day and be almost blind the next, but I still overreact to these fluctuations—exulting at the good times, despairing at the bad. "I librate between a glum and a frolic," as W. H. Auden put it in his poem "Talking to Myself."

I miss swimming terribly—the swimming pool is where I feel best, think best, and I need it every day. But I am not allowed to swim for two weeks following the surgery. Dr. Abramson knows well what a deprivation this is for me; he is a passionate swimmer, too—the walls of his office display various medals he has won. He might have been a professional athlete if he had not chosen medicine.

Not wanting to bother Dr. Abramson (though he said I should feel free to call him), I phoned Bob this morning to ask if he would check my eye. He came along with his ophthalmoscope, dilated the pupil, had a long, careful look, and then drew me a picture of what he saw: the melanoma like a black mountain in the middle of my retina; one side so steep, he said, it looked like "a cliff." He saw no signs of hemorrhage or anything amiss. But the blinding light of his ophthalmoscope caused me to lose all central vision in the eye for several hours. Whatever I looked at with my right eye disappeared—the center of my clock disappeared, leaving a halo of peripheral vision around it (I dubbed this, in my mind, "bagel vision"). It gave me a sense of horror. If this were permanent and if it affected both eyes, it would be terribly incapacitating—is this what people with macular degeneration have to live with?[1]

JANUARY 18, 2006

NOON: The eye was still quite blurred and dilated at nine o'clock this morning, but in the past three hours this has diminished, and the 12 and 1 are starting to be visible again when I fixate on the center of the clock.

But something has happened to the perception of color in the eye. When I went for a stroll this morning, a bright green tennis ball lying in the gutter lost all its color when I looked at it with just my right eye. Similarly with a Granny Smith apple and a banana—both turned a horrid gray. Holding the apple at arm's length, I find the central gray-out surrounded by a nor-

1. Many people with macular degeneration still manage to lead pretty full and independent lives. One patient of mine, a feisty old lady, told me that for five years after she lost central vision from macular degeneration, she "operated quite well on peripheral vision." She could still take walks and find her way around, even though she was legally blind, with vision of 20/200 or less.

mal green, as if color vision is preserved around my fovea but not in it. Blues, greens, mauves, and yellows all seem to be attenuated or lost; bright reds and oranges are the least affected, so when I pick an orange from my fruit bowl to test myself with, its color looks almost normal.

JANUARY 25, 2006

Today and yesterday, the twelfth and thirteenth days after the end of the radiation treatment, I observed, for the first time in a week, definite signs of improvement. Apples are starting to regain their greenness, and acuity too has improved. Last night I was able to read normal-sized print (Luria's autobiography) for half an hour before I went to sleep. I had not been able to read myself to sleep, my usual custom, for most of the month, since going into the hospital.

But strange dreams, sometimes nightmares, continue. In one, two nights ago, people were being tortured, blinded, by having red-hot needles thrust into their eyes. When it was my turn I struggled, let out a feeble cry, and tore myself into wakefulness. Yesterday I was awakened (or perhaps I was only half asleep) by lightning. I was surprised—no storms had been predicted—and waited for the thunder. No thunder. The sky was clear. I then realized that this had probably been a flash from my damaged and abnormally active retina. I had had scintillations before, and coruscation, but never a fulguration of this sort.

This morning I dreamt of a grove of tea trees, which, I understood, exerted a powerful protection against cancer if one lived beneath them.

JANUARY 26, 2006

It is only 8 a.m., and there are already nine people here in Dr. Abramson's waiting room. Do they, do we, all have ocular melanoma? There are no children today, but there are several youngish adults, of both sexes, though ocular melanoma is commoner after the age of sixty. Was I carrying an ocular melanoma gene at forty, or twenty? Or was it a mutation, one of the many, ever increasing on our polluted, carcinogenic planet?

I tell Dr. Abramson about the temporary loss of central vision in the right eye following the blinding light of Bob's ophthalmoscopy and the color changes I had noticed since. All this, he says, while perhaps exacerbated by the surgery, the radiation, the blinding light, is probably temporary and should disappear. Upon examination he sees a bit of necrosis and calcification of the tumor—the expected result of radiation. His impression: we are "on course," but I will probably need "touch-up" lasering in a month or so. I don't need to limit my activity anymore; I am free to swim. Hoorah!

7 P.M.: Despite everything, it has not been a wholly unproductive week. Kate has typed (and enlarged) two of my music chapters for me to go over, and I have seen several people with synesthesia this week, all fascinating in their different ways. Perhaps, in spite of my difficulties reading and my obsession with testing visual fields, color changes, etc., I can still hope to complete my music book.

For the next few weeks, I continued to experience fluctuations, with the right eye almost blind on some days and better on others, with "fish-eye" distortions and great sensitivity to light. I had to wear large, all-enveloping sunglasses outside and avoid dazzling sun or flashbulbs, which could blind me in that eye for hours. I wore a patch on the eye for much of the time, so that the normal image from my good, left eye would not have to compete with the distortions from the right eye. In March, Dr. Abramson followed up my radiation treatment with some lasering, and a couple of weeks later, the edema finally started to subside. With this, the vision in my right eye started to stabilize, the distortions and light sensitivity gradually disappearing.

Abnormalities in color perception, however, remained, although (unlike the distortions) these were not apparent if I used both eyes. If I closed my good eye, I suddenly found myself in a different chromatic world. A field of yellow dandelions would suddenly become a field of white dandelions, while darker flowers would turn black. A bright green fern, a selaginella, turned a deep indigo when I scrutinized it through a lens with my right eye. (My right eye was always the dominant one, and I would automatically hold a lens or monocular to that eye, even though it was now so much worse than the left.)

There were also curious suffusions or diffusions of color. When, for example, I looked with my right eye at a pale mauve flower surrounded by green leaves, the green surround took over and filled in, so that the whole flower appeared green. When I looked at a meadow of bluebells and closed my left eye, the bluebells turned green, no longer distinct from the vegetation around them. It was like a conjuring trick—now you see it, now you don't—and quite extraordinary to perceive such different worlds with each eye.

When I saw Dr. Abramson in May, he said that the edema had entirely gone and the tumor had started to shrink, and that with luck, I might hope to enjoy good and stable vision for years to come.

All continued well over the next two months, and I made fewer and fewer entries in the heavy black notebooks marked "Melanoma Journal." I did not resume detailed notes for almost a year. But starting in July 2006, there was a gradual return of visual problems—especially distortion, diminished acuity, and sensitivity to light—and some regrowth of one area of the tumor.

Dr. Abramson used the gentler word "persistence" to describe this and thought that another, milder lasering would take care of it. But when I had the procedure in December, it did not help. It began to look as if that narrow strip of retina next to the fovea, which he had carefully avoided lasering in order to maintain some central vision, would have to be sacrificed after all.

By April 2007, distortions had become extreme in the right eye, and this affected my vision even with both eyes open. People turned into bizarre, elongated, El Greco–like figures, tilted to the left—they made me think of the insectlike Selenites pictured in my edition of H. G. Wells's **The First Men in the Moon.** And the sort of visual spread which had started a year before, at first confined to colors, now affected everything I looked at. Faces, in particular, would develop translucent, puffy, almost protoplasmic protuberances, like a Francis Bacon portrait.

I found myself involuntarily closing my right eye more and more. Its acuity, by May of 2007, had plummeted to 20/600—I couldn't read the largest letter on the screen. Up to this point I had thought of losing central vision as a disaster,

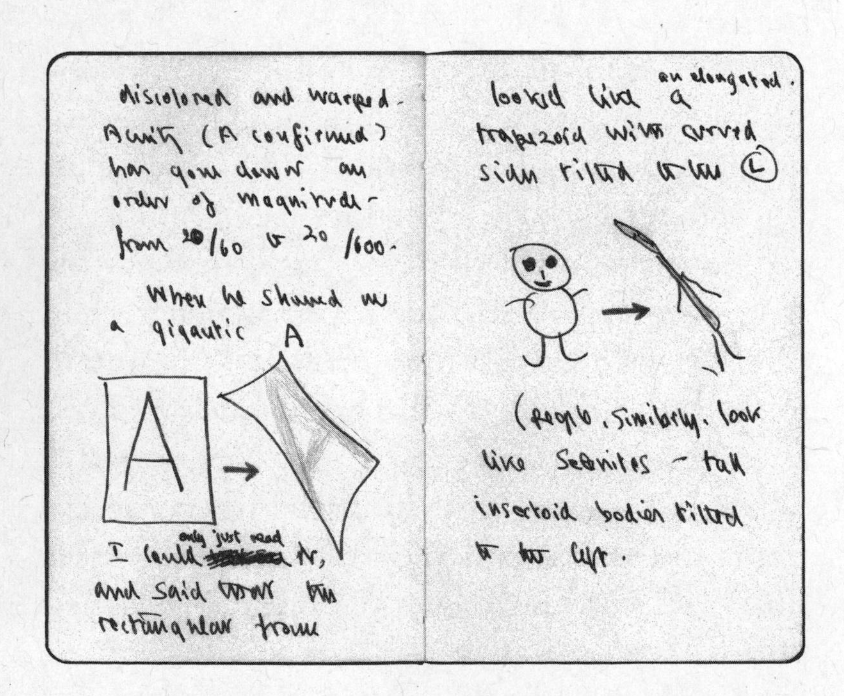

but now my sight was becoming so poor and so distorted that I began to wonder whether I would do better with no central vision in the right eye at all. Increasingly, it seemed, there was less and less to lose—so we scheduled a third lasering, which would finally knock out the rest of the tumor and, perhaps, the remaining central vision in that eye.

JUNE 2007

The lasering, a couple of weeks later, took about an hour, involving dozens of minute cauterizations, and I left the hospital with a heavy dressing over the eye, to protect it until the anesthesia wore off. Around 9 p.m. that night, I removed the dressing, not knowing what I would see, or not see.

I saw a huge black opacity partly obscuring central vision, like an amoeba with pseudopodia. It seemed to expand, contract, pulse—but its edge was razor-sharp. I stuck a finger into it, and the finger vanished, engulfed as by a black hole. Going to the bathroom mirror, facing my reflection, I could not see my own head with the right eye—only my shoulders and the bottom of my beard. I could not see the tip of the pen when I wrote.

When I went out the next morning, I saw only the lower halves of people walking. I was reminded of how, in Joyce's **Ulysses**, there is a Signor Artifoni who is characterized as "a pair of stout trousers" walking around Dublin. The streets were full of skirts and slacks, moving legs and hips with no upper halves. (A few days after this, the scotoma spread, and I could see only their feet.)

This, of course, is when I close the left eye. With both eyes, my vision is now remarkably "normal"—far more so than it has been for months, now that the right eye does not interfere with the left. It is out of the running, completely blind, at least so far as central vision goes. Oddly, this is a huge relief—I wish I had had it lasered months ago.

Stereo vision, however, now that I am mostly monocular, is quite compromised—completely missing in the upper half or two-thirds of my visual field, though partly intact in the bottom, where I retain some peripheral vision. So I see the lower halves of people in stereoscopic depth, while their upper halves are completely flat and two-dimensional. And, of course, as soon as I **look** at their lower halves, using what is left of my central vision, these become flat, too.

. . .

That first evening when I took off the bandage, I saw with my right eye a black blob, an amoeba. By the next day, this had settled into a darkness with the shape of Australia, complete with a little bulge in the southeast corner—I thought of this as its Tasmania. I was struck that first night by the fact that when I looked up at the ceiling, the blob disappeared, becoming so camouflaged that I could no longer be sure of its existence. I had to test to be sure, but it was still there—my black hole had become a white hole, had taken on the color of the ceiling around it. It was still a hole, and if I brought my finger from periphery to center, the finger would disappear as soon as it crossed the now-invisible margin of the scotoma.

I knew that the normal blind spot, which we all have, where the optic nerve enters the eye, is automatically filled in, so we are unaware of its existence. But the normal blind spot is tiny, whereas my own scotoma was huge, blotting out more than half of the entire visual field of the right eye. And yet, within a second or two of looking at a white surface, it could completely fill in, becoming white instead of black. The next day I tested this with a blue sky and found the same result. The scotoma became as blue as the sky, but this time I had no need to plot its margins with my finger, for when a flock of birds flew by, they suddenly disappeared into my scotoma, emerging on the other side a few seconds later—as if they had been cloaked in invisibility like a Klingon warship.

This filling in, I discovered, was strictly local, depending on a steady fixation of gaze. If there was a slight movement of the eye, the filling in dissipated, and the ugly black amoeba was back. Local, but persistent, for if I looked at a red surface for a few minutes and then at a white wall, I would see a large red amoeba (or Australia) on the wall, which would last about ten seconds before it turned white.

The blind spot, so called, does not just fill in color, it fills in patterns too, and I enjoyed experimenting with my own scotoma, testing its powers and limitations. It was easy to fill in a simple repetitive pattern—I started with the carpet in my office—though a pattern took a bit longer than a color, perhaps needing ten or fifteen seconds to duplicate. It would fill in from the edges, like ice crystallizing on a pond. The spatial frequency and fineness of detail in the pattern was crucial. My visual cortex had little trouble filling in fine-grained patterns, but coarser patterns were beyond it. So if, for instance, I stood two feet from a brick wall, my scotoma would turn brick red in color, but with no detail. If I stood twenty feet away, it would be filled in by perfectly respectable-looking brickwork.

Whether the brickwork was exactly the same as the original I could not be sure, but it was good enough to form a plausible simulacrum of the "missing" wall. I could be sure of exact replication only if I was gazing at absolutely predictable, repetitive patterns like chessboards or wallpaper. Once when I looked at a sky filled with fat woolly clouds, the pseudo-sky generated within the scotoma contained thin wispy clouds. I felt that my visual cortex was doing the best it could, perhaps by sampling or estimating the ratio of white cloud to blue sky, even though the actual shapes of the individual clouds were not right. I started to think of

my visual cortex not just as a rigid duplicating device, but as an averaging device, capable of sampling what was presented to it and making a statistically plausible (if not photographically accurate) representation of it. I wondered if this was what cuttlefish and octopuses did when they camouflaged themselves, taking on the colors, patterns, and even textures of the seafloor or plants or coral around them—not exactly, but plausibly enough to fool both predators and prey.

I found that movement could also be filled in to some extent. If I looked at the Hudson River, slowly swirling or rippling with small waves, these too were reproduced in my scotoma.

But there were strict limits. I could not simulate a face, a person, a complex object. I could not fill in my own head in the mirror when it was blanked out by my scotoma. And yet here I made another discovery, one that filled me with wonder. Idly playing, scotomizing, one day, I looked at my foot with my right eye and "amputated" it with my blind spot, a little above the ankle. But when I moved my foot a little, wiggling the toes, the stump seemed to grow a sort of translucent pink extension with a ghostly, protoplasmic halo around it. As I continued wiggling my toes, this took on a more definite form until, after a minute or so, I had a complete phantom foot, a visual phantom equipped with the missing toes, which seemed to move with the movements I was making. The foot did not look wholly solid or real, for it lacked surface detail, the appearance of skin—but it was very remarkable nonetheless. A similar thing occurred with my hand if I scotomized it, "amputated" it above the wrist. I subsequently tried to do the same with others' hands, but that did not work in the least. It was clear that my own foot or my own hand, my own movements and sensations, my own body image or intentions, were required.

After my lasering in June, I noticed that I could visualize my arms or other parts of my body in action, even when my eyes were closed, much more clearly and vividly than I had ever done before. "Seeing" my arms as I moved them seemed to attest to a heightened sensitivity or connection between the visual and the motor areas of the cortex—an intensity of communication or correlation between them unprecedented in my experience.

Another odd thing struck me within a day or two of the lasering in June of 2007. At one point, after gazing at the bookshelves in my bedroom for a few minutes, I closed both eyes and saw, for ten or fifteen seconds, the hundreds of books arrayed on the shelves in great, almost perceptual detail. This was not filling in but something quite different—a persistence of vision similar to what I had experienced in the hospital eighteen months earlier, when I seemed to see the washbasin so clearly "through" my eye patch.

Perhaps the loss of central vision in the right eye was equivalent to having it covered by a postoperative patch, in terms of depriving the brain of perceptual information. I had the sense that my visual cortex was now in a heightened or sensitized state, released to some extent from purely perceptual constraints.

Something similar happened a few days later, as I walked up to a crowded intersection full of bicycles, cars and buses, and people bustling in all directions. When I closed my eyes for a minute, I could still "see" the whole complex scene, full of color and movement, as clearly as if I had my eyes open.

I found this especially surprising, since I normally have very meager powers of visualization. I have difficulty evoking a mental picture of a friend's face, or my living room, or anything at all. The persistence of vision I had experienced was richly, mindlessly detailed, much more so than any voluntary image. It was so detailed I could see the colors of cars and sometimes read their license plates, to which I had paid no conscious attention. Involuntary, unselective, unstoppable, the image seemed to me akin to photographic or eidetic imagery—but, unlike eidetic imagery, it had a very definite and brief duration, lasting ten or fifteen seconds and then fading.

At one point, as I was walking with a friend, I saw two men walking towards us, both wearing white shirts, brilliantly clear in the late-afternoon sun. I stopped and closed my eyes, and found that I could continue to watch them, seemingly still walking towards us. When I opened my eyes, I was startled to find that the men in white shirts were nowhere to be seen. They had, of course, walked past us, but I was so engrossed in what I "saw" with my eyes closed—an arrested fragment of the past—that I got a sudden shock of discontinuity. I say "arrested," but what I saw in my mind's eye had motion, too. The men were walking, striding, yet they remained in the center of my mind's eye as they walked, without getting anywhere, as if on a treadmill. I had captured this bit of motion, like a film loop, which recycled in my mind even after they had gone. This had a paradoxical quality, like a snapshot of movement without any actual transit.

I rather enjoyed this persistence of vision, and Times Square, with its brilliant colored lights, its moving and flashing billboards, became a favorite place for testing it. The most potent stimulus of all was optic flow, a brisk stream of images past my eyes, which I could especially relish when I was a passenger in a fast-moving car.

I felt there was an analogy and perhaps a kinship between the filling-in phenomenon and the persistence of vision. Both came on strongly after the loss of central vision, though there had been intimations of both before. Both remained strong for two to three months in the summer of 2007 and then grew weaker (though they continue, in an attenuated form, to the present). "Filling in" seemed to me an inadequate term for a process which did not always confine itself to reconstituting a blind area but could go on to a sort of incontinent visual spread. (This, too, had been foreshadowed in those last, half-blind weeks before the June lasering, when faces spread and protuberated like monstrous, Francis Bacon faces.)

I experimented with this visual spread one day by gazing with my right eye at an old tree with a particularly exuberant and brilliantly green mass of foliage. Filling in soon occurred, so that the missing area turned green and textured to match the rest of the foliage. This was followed by a "filling out," an extension of the foliage, especially towards the left, resulting in a huge lopsided mass of "leaves." I realized how outlandish this had become only when I opened my left eye and saw the tree's actual shape. I went home and looked up an old paper by Macdonald Critchley on types of "visual perseveration" which he called "paliopsia" and "illusory visual spread."[2] Critchley saw these two phenomena as analogous: one a perseveration in time, the other a perseveration in space.[3]

---

2. Although Critchley coined the term "paliopsia," most people now use "palinopsia."

3. Frigyes Karinthy, in **Journey Round My Skull**, describes a very different sort of filling in when he is losing his vision. It is not the sort of low-level filling in that I have, but a much more complex filling in at a higher level, one that draws on association and memory:

> By now, I had learnt to interpret every hint afforded by the shifting of light and to complete the general effect from memory. I was getting used to this strange semi-darkness in which I lived, and I almost began to like it. I could still see the outline of figures pretty well, and my imagination supplied the details, like a painter filling an empty frame. I tried to form a picture of any face I saw in front of me by observing the person's voice and movements. People were often astonished to see that I could not distinguish between colours and shades, yet I would catch momentary facial expressions unnoticed by those with normal eyesight. I, too, was surprised. The idea that I might already have gone blind struck me with

Here perhaps one has to use the word "pathological," for one can hardly have a normal visual life if every perception gets extended and smeared in space and time; one needs restraint or inhibition, clear boundaries, to preserve the discreteness of perception.

Critchley's patients had brain tumors or other cerebral disorders, whereas I had only retinal damage. Yet clearly I was also experiencing cerebral phenomena—I supposed that retinal impairment had led to some abnormal excitation in my visual cortex. Many years ago—I described this in **A Leg to Stand On**—I had an injury to the nerves and muscles in one leg that caused some strange cerebral symptoms similar to those of a parietal lobe disorder. When I wrote to the Russian neuropsychologist A. R. Luria about this, he spoke of "the central resonances of a peripheral disorder." Now I was experiencing such a resonance in the realm of vision.

---

sudden terror. . . . I might only be using people's words and voices to reconstruct the lost world of reality, just as our mind, at the moment we fall asleep, forms images resembling those of real life from the phosphenes that dance before our closing eyes. I stood on the threshold of reality and imagination, and I began to doubt which was which. My bodily eye and my mind's eye were blending into one, and I could no longer be certain which of the two was really in control.

. . .

In June of 2007 I also had a sharp surge of hallucinations—apparitions that came out of the blue and had no relation to the external world—and this has continued, to some extent, ever since. Neurologists speak of simple or elementary visual hallucinations, as opposed to complex ones. In the simple ones, there are hallucinations of color, shapes, and patterns; in the complex ones, there may be figures, animals, faces, landscapes, etc. For the most part, I have simple ones.

Almost from the start, sparks, stripes, or blobs of light have appeared in my visual field, as well as complex patterns resembling alligator hide. I sometimes think that a wall is patterned or textured when it is not, and have to touch it to be certain whether the stippling I see is real.

I often see a multitude of little tufts, like tussocks of grass, all over my visual field, even with both eyes open. At other times there are checkerboards, usually black-and-white, but sometimes faintly colored. The apparent size of these checkerboards depends on where I am "projecting" them. If I look at a piece of paper six inches away, I might see a checkerboard on it the size of a postage stamp; if I look at the ceiling, it might appear to be a foot square; if I look at a white wall across the street, the checkerboard might be the size of a shop window. Some of my checkerboards are rectilinear, others are curvilinear, and some have an almost hyperbolic shape. Sometimes one checkerboard will undergo fission or multiplication, becoming a dozen smaller checkerboards, arrayed in rows and columns. Complex patchworks or mosaics are common, too, and seem to be variants or elaborations of the basic checkerboard patterns. These tend to switch from one to another in constant, kaleidoscopic change.

I also see tilings or tessellations composed of polygonal (often hexagonal) pieces, some flat and some three-dimensional, like honeycombs or radiolaria. Sometimes there are spirals or concentric rings, or radial patterns like filigreed doilies. Occasionally I see "maps"—maps of enormous, unknown cities, as might be seen at night from a low-flying airplane, with ring roads and radial spokes illuminated, looking like giant spiderwebs of light.

Many of these patterns are microscopically detailed. I see thousands of lights in my nocturnal cities. These images or hallucinations have greater clarity, are more fine-grained, than perception itself, as if my inner eye had an acuity of 20/5 rather than 20/20.

The most constant patterns (perfectly visible with both eyes open, and especially so if my visual field is otherwise blank) are sticklike or occasionally curved patterns resembling letters or numbers. Occasionally I recognize a 7 or a Y or a T or a delta, but for the most part they are unintelligible, like runes. They make me think of a child's letter box, with letters spilled out at random and at all angles. These are rather faint and often have double lines, giving the impression of being incised like the lettering on a stone. These pseudo-letters and pseudo-

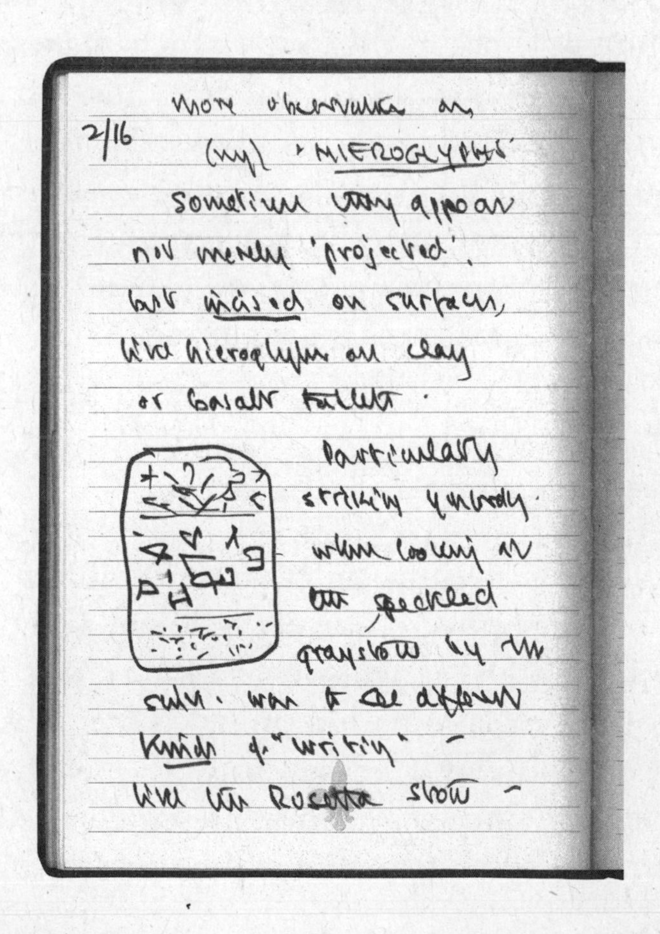

numbers often flicker and form, dissolve and re-form in fractions of a second, all over my visual field. Sometimes, if I am looking at a horizontal segment of a wall, the runes come in a row, like a frieze.

Most of the time, I am able to ignore them, as I ignore the tinnitus I have had for the past few years. But often in the evening, when the sights and sounds of the day lessen, I may become suddenly aware of these faint hallucinations. And often it is a visual emptiness—a ceiling, a white washbasin, the sky—which makes me conscious of the visual patterns and images continually chasing across my visual field. Yet these little hallucinations are interesting, in a way: they show me the background activity, the idling, of my visual system, generating and transforming patterns, never at rest.

THURSDAY, DECEMBER 20, 2007

I had been feeling fairly relaxed about my tumor—it seemed relatively indolent and contained, and Dr. Abramson had said it was rare for ocular melanoma like mine to metastasize. But on Monday (the seventeenth, two years to the day since my tumor presented itself) I observed, in the gym, a roughly circular black spot the size of a dime on the skin just below my left shoulder. I was startled and frightened: the spot was jet black in color, with a clear border, and slightly raised; it looked nothing like an ordinary bruise. Was it, more ominously, the start of a skin melanoma, metastasized from the tumor in my eye?

When I showed the spot to Mark and Peter, who came for dinner tonight, both of them looked startled and worried. "It looks bad, very dark," Mark said. "I think you should get it checked within twenty-four hours." It did not, he added, look like a melanoma; but it did not look like anything he had ever seen before. Christmas holidays are upon us, as they were in '05, and this means that I should get it looked at tomorrow; otherwise it may have to wait until the New Year. I fear I will obsess about it, get myself into a near panic, if this cannot be clarified straightaway. I feel agitated now . . . I think I may have to sedate myself.

FRIDAY, DECEMBER 21, 2007

The dermatologist, Dr. Bickers, a kind, sensitive man, very knowledgeable, too—realizing my anxiety, worked me into his schedule today. He looked at my arm and the rest of my skin, saw nothing amiss. The blackness, he said, was just a little bleeding into one of the brown spots that increasingly mottle one's skin with age. Probably I had bumped into something; the blood would clear in a couple of days. I am greatly relieved—I would have gone mad waiting until January to have it checked.

For a decade or so before my melanoma, I had been an active member of the New York Stereoscopic Society; I had enjoyed playing with stereoscopes and stereo illusions since childhood. Seeing the world in depth always seemed as natural, as integral to my visual world as color vision. It gave me a sense of the solidity of objects and the reality of space—the wonderful, transparent medium in which they resided. I was keenly conscious of how my visual world instantly collapsed if I closed one eye and reexpanded the moment I opened it again. Like many of my fellow members of the Stereoscopic Society, I seemed to live in a deeper world, visually, than most people.

My experiences with Stereo Sue and her lyrical delight when she gained stereo vision after a lifetime of stereo blindness reinforced my feeling of appreciation for stereo vision. Indeed, I had spent much of 2004 and 2005 preoccupied with stereo vision, thinking and writing about it, and corresponding with Sue.

And then, in June of 2007, when the melanoma encroached on my fovea and had to be lasered, I lost all the central vision in that eye and, with it, stereoscopy. The complete and sudden flattening of the visual world I had experimented with as a boy by closing one eye now became a permanent condition. Some people have little stereo vision to begin with, or else they make so little use of binocular cues that they scarcely notice the difference if they lose stereoscopy. My situation was very different. Stereoscopy had been a central part of my visual life, and its loss had a profound impact at many levels, from the practical challenges of daily life to the whole concept of "space." Indeed, these changes were so radical that I was slow to recognize them fully.

Stereoscopy is most important in one's immediate vicinity, and it was here that I had initial problems of all sorts, some comic and some dangerous. When I reached out my hand for a canapé at a cocktail party, I might find myself grasping at air, missing the mark by six inches or more. I once poured wine into a friend's lap, missing the glass by almost a foot.

More dangerously, I fail to see steps or curbs, and may trip or come down with a jolt. If there are no shadows or ancillary cues, I see steps only as lines on the ground and have no idea how deep they are, let alone whether they go up or down. Particularly treacherous are those I cannot anticipate, such as a couple of steps in an outdoor plaza or in someone's sunken living room (often these also lack railings, which can serve as a visual cue). Going down a flight of stairs is a real and sometimes terrifying hazard, and I need to feel my way cautiously, testing for each step with my foot. Sometimes there may be such a compelling sense of flatness to my eye that it competes with what my foot says. Even when every other sense, including common sense, tells me that there is another step, if I cannot see its depth, I hesitate, confounded. After a long pause, I will trust my foot, but the dominating power of vision makes this far from easy.

These experiences (like many others in the past two years) set me to thinking of Edwin Abbott's classic 1884 book **Flatland**, where the inhabitants of his two-dimensional world are themselves two-dimensional geometric figures. Occasionally they are confronted by spontaneous changes in the appearance of things which can only be explained, their theorists tell them, if one postulates the existence of three-dimensional objects moving in a three-dimensional space, presenting slices of themselves as they intersect the plane of Flatland. Thus Flatlanders infer the existence of a spatial dimension they cannot see. This is a far-fetched analogy to my own situation, but it always comes to mind when I have to infer depth, despite the at times overwhelming flatness presented to my eye.

Paradoxically, I have lost my fear of heights. I used to feel a sort of frisson, a slight sense of alarm, when I looked down from a tall building to the street below. When I lived in Topanga Canyon, I would avoid getting near the precipitous edges of the winding canyon road. The thought of falling gave me the chills. But now that I have lost depth perception, these feelings have disappeared, and I can look down from great heights with complete indifference.

Occasionally, I have pseudo-stereo experiences, as when something flat, like a newspaper lying on the floor, looks to me as if it is sticking up in the air. Opening my door, I have mistaken my doormat for a table and come to an abrupt, confounded halt. Sometimes I imagine that there may be steps when I see lines on the ground, the edge of a rug, or some other boundary. Does the boundary go with a step, or not? So I have to stop and carefully test with a toe. I rarely experienced such misperceptions when I had both eyes, for stereoscopy serves to clarify and disambiguate situations where monocular cues may be ambiguous or deceiving.

Crossing streets, dealing with steps, just walking around—things that needed no conscious attention before—now require constant care and forethought. People who have spent most of their lives without stereo vision, like Sue, may adapt relatively easily to these challenges, but having been exceptionally and perhaps excessively biased to binocular cues for stereoscopy, I was finding it extremely difficult to function without two eyes.

I wake every morning to a cluttered world, everything on top of everything else. There is no room anywhere, no space between things.

I used to enjoy the tiny light bulbs strung on city trees at Christmas—they seemed to create globes of twinkling lights suspended in midair. Now I see a tree full of such lights as a disk, with no more depth than a sky full of stars. And when I go to the botanic garden, I can no longer gaze, as I once loved to do, at the thick foliage of trees and bushes and see layer after layer, depth upon depth—it is all a flat confusion now.

My reflection in the mirror no longer seems to be behind the mirror; it appears on the same plane as the mirror's surface. I see spots on my clothes in the mirror and try to brush them off, only to realize that they are spots on the surface of the mirror itself. A similar confusion made me think, one February day, that it was snowing inside the kitchen—"outside" the window seemed no farther away than "inside."[4]

While, for the most part, I hate the flatness of everything and lament the loss of depth, I occasionally have a sense of appreciation for my two-dimensional world. Sometimes I see a room, a quiet street, or a laid table as a still life, a beautiful visual composition, as I imagine it might be seen by a painter or a photographer

---

4. There have, however, been two incidents that I am hard put to explain. I had smoked a little cannabis both times, and I found myself totally absorbed, gazing in a sort of rapture at flowers—some narcissus in a pot on one occasion and morning glories twining up a fence on the other. Both times, it seemed to me, the flowers filled out before my eyes, thrust themselves into the space around them, assuming their full and proper three-dimensional glory. They deflated once again when the cannabis wore off. Was this vision "real" or an illusion? It was wholly different in quality from the pseudostereo, the confounding illusions of depth and distance I would sometimes have with lines on the ground, where there was not, in reality, any depth at all. The flowers did have depth, and I was seeing them as I used to when I had two good eyes. If it was an aberrant perception or an illusion, it was a veridical one, consonant with reality.

Some of my correspondents have occasionally experienced an opposite effect with cannabis—a loss of stereo, so that their visual world seems two-dimensional, like a painting.

constrained to a flat canvas or film. I find a new pleasure in looking at paintings or photographs, now that I am more conscious of the art of composition. They can be more beautiful in this sense, although they no longer yield me even the illusion of depth.

One afternoon I went to a nearby Japanese restaurant for sushi—one of the attractions of my sidewalk table being the sight of a ginkgo tree across the road. In the middle of the day, at that time of year, the sun's rays would cast a detailed shadow of the tree and its delicate leaves onto the yellow wall about five feet behind it. But, without stereoscopy, I now saw the tree and its shadow on the same plane, as if both were painted on the wall—a vision both alarming and exquisite, for the 3-D reality had turned into a Japanese painting.

Stereo vision at a distance may be less immediately important, but not being able to judge distance opens me to deep and often absurd doubts and illusions. In Edgar Allan Poe's story "The Sphinx," the narrator sees a gigantic jointed creature climbing a faraway hillside; only later does he realize that what he is seeing is a minute insect, practically in front of his nose. I found "The Sphinx" a little far-fetched until I lost stereoscopy. Now I have such experiences constantly. The other day I saw a piece of lint on my glasses and tried to clean it off, only to realize that the "lint" was a leaf on the sidewalk.

It is not just the sense of depth and distance which is undermined but, occasionally, the sense of perspective itself, so crucial to the recognition that one is in a world of solid objects arrayed in space. When I visited a friend's barn on Long Island, I first failed to recognize it as a barn, for I saw only vertical, horizontal, and diagonal lines, like a geometrical diagram inscribed on the sky. Then, all of a sudden, it acquired perspective and became recognizable as a barn, though still flat, like a photograph or a painting.

My inability to see depth or distance leads me to combine or conflate near and far objects into strange hybrids or chimeras. One day I was puzzled to find a gray web between my fingers, before I realized that I was seeing the gray carpet three feet below—now seen on the same plane as my hands and construed as part of them. I was horrified once, looking at a friend in profile, to notice twigs or slivers of wood coming out of her eyes—but these belonged, I soon realized, to a tree across the road. I spotted a man crossing the road in Union Square, with an enormous scaffolding on his shoulders—he is crazy to be carrying such a thing,

I thought—and then I realized that the scaffolding was thirty feet behind him, another conflation. Another time I saw the top of a fire engine apparently impaled on the roof of my car, and then realized that the fire engine was a dozen yards behind the car. But knowing this, or moving my head to demonstrate it by motion parallax, makes strangely little difference to the illusion.

A gigantic pontoon a hundred feet high, spotted at a traffic jam, turns out to be the side-view mirror of a car just in front of me. A woman's strange green umbrella turns out to be a tree a hundred feet behind her. Most frighteningly, as I was reading in bed one night, I "saw" the ceiling fan about to crash into the reading light just above my head—I "know" that the two objects are at least four feet apart, but this did not prevent the sudden illusion.

Nothing protrudes or recedes from me anymore; there is no direct sensation of "before" or "behind," only an inference based on occlusion and perspective. Space was once a hospitable, deep realm in which I could locate myself and wander at will. I could enter it, I lived in it, I had a spatial relationship to everything I could see. That sort of space no longer exists for me visually—or mentally.

After two years without stereoscopy, I now

function pretty well. I have learned how to shake hands, pour wine, and negotiate steps. I have returned to bicycling and driving my car—these are made possible by motion parallax and the fact that perception is being complemented by action, that I am **acting** in a three-dimensional world, even though it still looks two-dimensional to me. Most of the time, I can "see through" my illusions and conflations. But this does not alter my sense that a fundamental aspect of the visual world has been taken away and that things will never look as they did before, will never look right. The visual reality I face is utterly wrong—for I know so well how things used to be, and should be.

The only time I see in stereo now is in dreams, for I have had occasional stereo dreams all my life—usually dreams in which I am looking through a stereoscope at an exquisite pair of stereo photographs, perhaps at an urban landscape or the depths of the Grand Canyon. I wake from these to a reality which is incorrigibly, irreversibly, maddeningly flat.

My vision remained in this state, fairly stable, for two years. I was able to do most of the things I wanted—for having peripheral vision in my right eye still allowed me a full visual field, even if it lacked straight-ahead depth. With this peripheral vision, I maintained a small crescent of stereopsis near the bottom of the visual field, and this was important in giving me some implicit or unconscious sense of depth and space, even if there was no stereopsis in the rest of the visual field. But it could be very tantalizing, too, for the region of stereopsis lay below my fixation point, and whenever I tried to focus on something with my one good eye, it immediately flattened out.[5]

All this was to change on September 27, 2009. The day opened like any other; I had my swim, ate breakfast, and was cleaning my teeth, when it seemed to me that a film came over my right eye. Its peripheral vision, the only vision it still had, was hazy. I wondered if my glasses had misted over, so I took them off and cleaned

5. Gradually the peripheral vision in my right eye got worse, as a cataract developed in reaction to the radiation treatment. With this, what little stereopsis I had diminished. When I had the cataract removed in the spring of 2009, I had a sudden resurgence of peripheral vision and stereopsis. Everything looked, with my right eye, brigher and bluer, and when I went to the orchid show at the botanical garden the next day, I not only saw colors with a startling brilliance and freshness, but I could see flowers thrusting towards me in the bottom of my visual field. I rejoiced in this, but did not realize how short-lived my return to (at least partial) stereo would be.

them—but the film was still there. I could see objects through it, but their outlines were indistinct.

"One of those things," I thought (although it was unlike anything I had ever experienced before). "It will clear in a few minutes." But it did not clear. It grew denser and denser. A feeling of fear and danger took hold of me—what was going on? I phoned Dr. Abramson's office; he was away, but his colleague suggested that I come to the office straightaway. Looking into the eye, Dr. Marr confirmed my suspicion: there was bleeding, probably from the retina, and the blood was now seeping into the vitreous humor at the back of the eye. The cause of the hemorrhage was unclear, but the tumor, irradiation, and repeated laserings might well have scarred the retina, making it more fragile, increasing the chances of a blood vessel being eroded or giving way. There was nothing to be done at this point.

By late afternoon I could not count my own fingers or see anything distinctly with the right eye. I could only sense diffuse illumination from the window and some movement, the way one can see, in bright light, a hand waving in front of the eyes even when the eyelids are closed. The blood would eventually clear, I was told, but this could take six months or more—

now, for all practical purposes, my right eye was completely blind.

I could not help thinking of that other day, the day everything started to go wrong, at the end of 2005—and of the nearly four-year fight in which the eye carried on, with ever more of the retina being nibbled at or blasted away. Was this the final knockout blow?

To take my mind off things visual, I went to the piano, closed my eyes, and played for a while. Then, to dull my feelings and prevent rumination, I took two sleeping tablets and went to bed.

I slept deeply. Awakened by my clock radio, I listened with my eyes closed, in that dreamy state between wake and sleep, and it was only when I opened my eyes and saw nothing with the right eye but a vague dim light where the morning sun was flooding in that the memory of what had happened suddenly came back to me.

On Monday morning, Kate came over and suggested that we go for a walk together. As soon as we emerged into the morning bustle of Greenwich Avenue, crowded with people balancing coffee cups and cell phones, people walking dogs, parents with children going to school, I realized that I was in trouble. I was startled, even terrified, because people and objects suddenly seemed to materialize, to loom at me on the right side without any warning. Had Kate not been walking on my right, protecting my blind side, I would have been colliding with everything, tripping over dogs, crashing into strollers, without the least awareness that they were there.

We do not honor our peripheral vision as much as we should, for most of the time we have little explicit consciousness of it. We look, we fixate, we target with our foveas, our central vision. But it is peripheral vision, surrounding this, which gives us a context, a sense of how whatever we are looking at is situated in the wider world. And it is especially movement that peripheral vision is tuned to: peripheral vision alerts us to unexpected movements on either side, and then central vision moves in to target these.

For me, now, a biggish slice of the periphery to my right—forty degrees or more, like a very large slice of cake—has been carved out of my vision. I see, roughly speaking, nothing to the right side of my nose.[6] I had lost central

---

6. There may be various optical or mechanical ways to enlarge the visual field if an eye is lost. The use of a prism, for example, may allow an extra six or eight degrees of visual field, and there may be ingenious strategies with mirrors, too. A more drastic solution was attempted by Federico, a fifteenth-

vision in the eye earlier, but I still had enough peripheral vision to give me a forewarning, an intimation, of things happening on that side. But now I have lost even this. I have no awareness here, and whatever comes into my visual field from that side is unexpected and startling. I cannot overcome the sense of bewilderment, even shock, when people or objects appear suddenly to my right. A massive slice of space no longer exists for me, and the idea that there **could** be anything in that space has likewise disappeared.

Neurologists talk of "unilateral neglect" or "hemi-inattention," but these technical terms do not convey how outlandish such a state can be. Years ago, I had a patient with a startling neglect of her own left side, and the left side of space, due to a stroke in her right parietal lobe.[7]

---

century duke of Urbino who had lost one eye in a tournament. Fearing the ever-present threat of assassination and wanting to preserve his prowess on the battlefield, he had his surgeons amputate the bridge of his nose to allow a wider field for the remaining eye.

7. I wrote about this patient in "Eyes Right!" in **The Man Who Mistook His Wife for a Hat.** Another example was provided by my colleague M.-Marsel Mesulam, who wrote, "When the neglect is severe, the patient may behave almost as if one half of the universe had abruptly ceased to exist in any meaningful form. . . . Patients with unilateral neglect behave not only as if nothing were actually happening in the left hemispace, but

But this had not prepared me at all to find myself in a virtually identical situation (though caused, of course, not by a cerebral problem but by an ocular one). This came home even more forcefully when Kate and I finished our walk and headed back to my office. I walked ahead and got into the elevator—but Kate had vanished. I presumed she was talking to the doorman or checking the mail, and waited for her to catch up. Then a voice to my right—her voice—said, "What are we waiting for?" I was dumbfounded—not just that I had failed to see her to my right, but that I had even failed to imagine her being there, because "there" did not exist for me. "Out of sight, out of mind" is literally true in such a situation.

NOVEMBER 9, 2009

Six weeks have passed now since the hemorrhage. I had expected, in time, to accommodate to my semiblindness, my hemispace, but that has not happened. Every time someone or something suddenly appears to my right, it is just as unexpected as the first time. I am still

---

also as if nothing of any importance could be expected to occur there."

in a world of suddenness and discontinuity, of sudden apparitions and disappearances.[8]

I can deal with this only by constantly turning my head to monitor what is going on in the blind area. (Indeed, I have to twist my entire upper body around to compensate for the sixty degrees or so I am missing.) But doing this is not only wearisome, it feels absurd, because so far as my own perception is concerned, I have a full visual field—nothing is missing for me, subjectively, and so there is nothing to look for. It may appear odd to other people, too, who feel that I am acting bizarrely, contorting my body or turning around and staring at them.

---

8. John Hull, who became totally blind in midlife, describes this sense of suddenness in **Touching the Rock:**

> For the blind, people are not there unless they speak. Many times I have continued a conversation with a sighted friend only to discover that he is not there. He may have walked away without telling me. He may have nodded or smiled, thinking the conversation was over. From my point of view, he has suddenly vanished.
>
> When you are blind, a hand suddenly grabs you. A voice suddenly addresses you. There is no anticipation or preparation. . . . I am passive in the presence of that which accosts me. . . . The normal person can choose whom he wants to speak to, as he wanders around the streets or the market-place. People are already there for him; they have a presence prior to his greeting them. . . . For the blind person, people are in motion, they are temporal, they come and they go. They come out of nothing; they disappear.

There are parallel experiences with senses other than vision. If one has a complete spinal anesthesia, for instance, one loses all sensation and power of movement in the lower half of the body. But this description does not do full justice to the strangeness one can encounter. One's awareness, the sense of one's body, is sharply terminated, in effect, at the level of the anesthetic, and what lies below is no longer felt as part of oneself, for it is not sending any information to the brain testifying to its own existence. It has disappeared, taking its place, its space with it.

One can, of course, **look** at one's "missing" legs, and this is even stranger in a way, for the legs seem curiously unreal, alien—almost like wax models from an anatomy museum. It has been shown, with functional imaging, that the anesthetized parts of the body actually lose their representation in the sensory cortex. So it seems with the right side of my visual field—it no longer sends any signals to the brain, no longer has any representation there. As far as the brain is concerned, it does not exist.

DECEMBER 6, 2009

It is now ten weeks since my hemorrhage, and I have still achieved surprisingly little in the way of accommodation. I must remind myself again and again to check, to make sure that nothing on the blind side is being ignored or forgotten—it is still far from automatic. I wonder if I will ever accommodate, and I think of something one of my correspondents, Stephen Fox, wrote:

> Far worse than the loss of depth was the new limitation on the visual field. My right arm became covered in bruises from running into door frames because my brain was still reacting as if it was getting the full panorama from two eyes. I also often knocked objects off the table with my right arm. In fact, limited scope remains a problem even after 22 years, especially in crowded subway stations where people's paths may suddenly and silently converge on my right, resulting in occasional, embarrassing collisions.

Greenwich Avenue, and the outside world generally, remain as full of hazards, real and imaginary, as when I went for that first post-hemorrhage walk many weeks ago. People rush around, so preoccupied with cell phones and text messaging that they themselves are functionally deaf and blind, oblivious of their environment; others have tiny, insectlike dogs on long, invisible leashes, which function as trip wires for the visually impaired; children zip around beneath eye level on scooters. There are other dangers, too: manholes, grates, and fire hydrants, doors suddenly opening, cyclists delivering lunches—the entire scene seems designed to drum up business for orthopedists. I dare not walk alone; fortunately, my friends help out by walking with me, acting as guides and protectors on my blind side. And I would not dream of driving at this point.

I try to stick to the right side of the sidewalk so that no one can overtake me on my blind side, but this is not always possible: the sidewalk is often crowded and not mine to commandeer as I might like. I find myself losing things on my own desk—my reading glasses, my fountain pen, a letter I have just written—if I have placed them to my right.

And yet (so I am told, in Frank Brady's book **A Singular View: The Art of Seeing with One Eye**) almost everyone who loses an eye does accommodate to its loss, more easily if they are young or if the loss of vision is gradual; especially too if the affected eye is not the dominant one and vision in the other eye is good. (I, alas, rank rather low on all these criteria.) Most people, given time, are able to return to a full and free life—so long, Brady emphasizes, as they retain a special mindfulness, a hyperconsciousness, of the missing side.

Perhaps this will be possible for me, too, in the future. But it is far from my situation now. Strange incidents seem to beset me all the while. Returning the other day from a walk with my friend Billy, I "lost" him when I entered the elevator. I turned to the right and someone was standing there, someone who I thought for a moment must be Billy. Then I realized that it was a stranger, a stranger who himself looked surprised and puzzled, even slightly alarmed, at my turning and staring at him with a look of befuddlement. He must have thought I was mad. It was only when I twisted still farther to the right that I found Billy, to the left of the stranger, deep in my nowhere.

Five minutes later, when we got to my apartment and I turned to put on a kettle for tea, Billy vanished again—but I discovered him, after a bewildered pause, precisely where I had left him. He had not moved, but my turning away had put him in my blind spot, my visual and mental "nowhere." I was again astounded that this could happen within seconds, and in a way so contrary to memory and common sense. Each time this happens, it is just as startling.

Time will tell whether I am able to adapt to this new visual challenge—or perhaps the hemorrhage will clear first and restore at least some peripheral vision to my right eye. In the meantime, I have a large "nowhere" in my right visual field and my brain, a nowhere of which I am not and can never be directly conscious. For me, people and objects continue to "disappear into thin air" or "come out of the blue"—these are no longer just metaphors for me, but as close as I can come to describing the experience of nothingness and nowhere.

The Mind's Eye

To WHAT EXTENT are we the authors, the creators, of our own experiences? How much are these predetermined by the brains or senses we are born with, and to what extent do we shape our brains through experience? The effects of a profound perceptual deprivation such as blindness may cast an unexpected light on these questions. Going blind, especially later in life, presents one with a huge, potentially overwhelming challenge: to find a new way of living, of ordering one's world, when the old way has been destroyed.

In 1990, I was sent an extraordinary book called **Touching the Rock: An Experience of Blindness,** by John Hull, a professor of religious education in England. Hull had grown up partly sighted, developing cataracts at the age of thirteen and becoming completely blind in his left eye four years later. Vision in his right eye remained reasonable until he was thirty-five or so, but there followed a decade of steadily failing vision, so that Hull needed stronger and stronger magnifying glasses and had to write with thicker and thicker pens. In 1983, at the age of forty-eight, he became completely blind.

Touching the Rock is the journal he dictated in the three years that followed. It is full of piercing insights about his transition to life as a blind person, but most striking for me was his description of how, after he became blind, he experienced a gradual attenuation of visual imagery and memory, and finally a virtual extinction of them (except in dreams)—a state that he called "deep blindness."

By this, Hull meant not only a loss of visual images and memories but a loss of the very **idea** of seeing, so that even concepts like "here," "there," and "facing" seemed to lose meaning for him. The sense of objects having appearances, or visible characteristics, vanished. He could no longer imagine how the numeral 3 looked unless he traced it in the air with his finger. He could construct a **motor** image of a 3, but not a visual one.

At first Hull was greatly distressed by this: he could no longer conjure up the faces of his wife or children, or of familiar and loved landscapes and places. But he then came to accept it with remarkable equanimity, regarding it as a natural response to losing his sight. Indeed, he seemed to feel that the loss of visual imagery was a prerequisite for the full development, the heightening, of his other senses.

Two years after becoming completely blind, Hull had apparently become so nonvisual in his imagery and memory as to resemble someone who had been blind from birth. In a profoundly religious way, and in language sometimes reminiscent of that of Saint John of the Cross, Hull entered into the state of deep blindness, surrendered himself, with a sort of acquiescence and joy. He spoke of deep blindness as "an authentic and autonomous world, a place of its own. . . . Being a whole-body seer is to be in one of the concentrated human conditions."

Being a "whole-body seer," for Hull, meant shifting his attention, his center of gravity, to the other senses, and these senses assumed a new richness and power. Thus he wrote of how the sound of rain, never before accorded much attention, could delineate a whole landscape for him, for its sound on the garden path was different from its sound as it drummed on the lawn, or on the bushes in his garden, or on the fence dividing the garden from the road:

> Rain has a way of bringing out the contours of everything; it throws a coloured blanket

> over previously invisible things; instead of an intermittent and thus fragmented world, the steadily falling rain creates continuity of acoustic experience . . . presents the fullness of an entire situation all at once . . . gives a sense of perspective and of the actual relationships of one part of the world to another.

With his new intensity of auditory experience (or attention), along with the sharpening of his other senses, Hull came to feel a sense of intimacy with nature, an intensity of being-in-the-world, beyond anything he had known when he was sighted. Blindness became for him "a dark, paradoxical gift." This was not just "compensation," he emphasized, but a whole new order, a new mode of human being. With this, he extricated himself from visual nostalgia, from the strain or falsity of trying to pass as "normal," and found a new focus, a new freedom and identity. His teaching at the university expanded, became more fluent; his writing became stronger and deeper; he became intellectually and spiritually bolder, more confident. He felt he was on solid ground at last.[1]

---

1. Despite an initially overwhelming sense of despair on losing their sight, some people, like Hull, have found their full creative strength and identity on the other side of blindness. One thinks especially of John Milton, who started to lose his sight around the age of thirty (probably from glaucoma), but

Hull's description seemed to me an astounding example of how an individual deprived of one form of perception could totally reshape himself to a new center, a new perceptual identity. Yet I found it extraordinary that such an annihilation of visual memory as he described could happen to an adult with decades of rich and significant visual experience to call upon. I could not, however, doubt the authenticity of Hull's account, which he related with the most scrupulous care and lucidity.

Cognitive neuroscientists have known for the past few decades that the brain is far less hardwired than was once thought. Helen Neville was one of the pioneers here, showing that in prelingually deaf people (that is, those who had been born deaf or become deaf before

---

produced his greatest poetry after becoming completely blind a dozen years later. He meditated on blindness, how an inward sight may come in place of outward sight, in **Paradise Lost**, in **Samson Agonistes**, and—most directly—in letters to friends and in a very personal sonnet, "On His Blindness." Jorge Luis Borges, another poet who became blind, wrote about the varied and paradoxical effects of his own blindness; he also wondered how it might have been for Homer, who, Borges imagined, lost the world of sight but gained a much deeper sense of time and, with this, a matchless epic power. (This is beautifully discussed by J. T. Fraser in his 1989 foreword for the Braille edition of **Time, the Familiar Stranger.**)

the age of two or so) the auditory parts of the brain did not degenerate. They remained active and functional, but with an activity and a function that were new: they were transformed, "reallocated," in Neville's term, for processing visual language. Comparable studies in those born blind, or blinded early, show that some areas of the visual cortex may be reallocated and used to process sound and touch.

With this reallocation of parts of the visual cortex, hearing, touch, and other senses in the blind can take on a hyperacuity that perhaps no sighted person can imagine. Bernard Morin, the mathematician who showed in the 1960s how a sphere could be turned inside out, became blind at the age of six, from glaucoma. He felt that his mathematical achievement required a special sort of spatial sense—a haptic perception and imagination beyond anything a sighted mathematician was likely to have. And a similar sort of spatial or tactile giftedness has been central to the work of Geerat Vermeij, a conchologist who has delineated many new species of mollusks, based on tiny variations in the shapes and contours of their shells. Vermeij has been blind since the age of three.[2]

---

2. In his book **The Invention of Clouds**, Richard Hamblyn recounts how Luke Howard, the nineteenth-century chemist who first classified clouds, corresponded with many other

Faced with such findings and reports, neuroscientists began to concede in the 1970s that there might be a certain flexibility or plasticity in the brain, at least in the first couple of years of life. But when this critical period was over, it was thought, the brain became much less plastic.

Yet the brain remains capable of making radical shifts in response to sensory deprivation. In 2008, Lotfi Merabet, Alvaro Pascual-Leone, and their colleagues showed that, even in sighted adults, as little as five days of being blindfolded produced marked shifts to nonvisual forms of behavior and cognition, and they demonstrated the physiological changes in the brain that went along with this. (They feel it is important to distinguish between such rapid and reversible changes, which seem to make use of preexisting but latent intersensory connections, and the long-lasting changes that occur

---

naturalists of the time, including John Gough, a mathematician blinded by smallpox at the age of two. Gough, Hamblyn writes, "was a noted botanist, having taught himself the entire Linnean system by touch. He was also a master of the fields of mathematics, zoology and scoteography—the art of writing in the dark." (Hamblyn adds that Gough "might also have become an accomplished musician had his father, a stern Quaker . . . not stopped him playing on the godless violin that an itinerant fiddler had given him.")

especially in response to early or congenital blindness, which may entail major reorganizations of cortical circuitry.)

Apparently Hull's visual cortex, even in adulthood, had adapted to a loss of visual input by taking over other sensory functions—hearing, touch, smell—while relinquishing the power of visual imagery. I assumed that Hull's experience was typical of acquired blindness, the response, sooner or later, of everyone who loses sight—and a brilliant example of cortical plasticity.

Yet when I came to publish an essay on Hull's book in 1991, I was taken aback to receive a number of letters from blind people, letters that were often somewhat puzzled and occasionally indignant in tone. Many of these people wrote that they could not identify with Hull's experience and said that they themselves, even decades after losing their sight, had never lost their visual images or memories. One woman, who had lost her sight at fifteen, wrote:

> Even though I am totally blind . . . I consider myself a very visual person. I still "see" objects in front of me. As I am typing now I can see my hands on the keyboard. . . . I don't feel comfortable in a new environment

> until I have a mental picture of its appearance. I need a mental map for my independent moving, too.

Had I been wrong, or at least one-sided, in accepting Hull's experience as a typical response to blindness? Had I been guilty of emphasizing one mode of response too strongly, oblivious to other, radically different possibilities?

This feeling came to a head a few years later, when I received a letter from an Australian psychologist named Zoltan Torey. Torey wrote to me not about blindness but about a book he had written on the brain-mind problem and the nature of consciousness. In his letter he also mentioned that he had been blinded in an accident at the age of twenty-one. But although he was "advised to switch from a visual to an auditory mode of adjustment," he had moved in the opposite direction, resolving to develop instead his inner eye, his powers of visual imagery, to their greatest possible extent.

In this, he said, he had been extremely successful, developing a remarkable power of generating, holding, and manipulating images in his mind, so much so that he had been able to construct a virtual visual world that seemed as real and intense to him as the perceptual one he had lost—indeed, sometimes more real, more intense. This imagery, moreover, enabled him to do things that might have seemed scarcely possible for a blind man.

"I replaced the entire roof guttering of my multi-gabled home single-handed," he wrote, "and solely on the strength of the accurate and well-focused manipulation of my now totally pliable and responsive mental space." Torey later expanded on this episode, mentioning the great alarm of his neighbors at seeing a blind man alone on the roof of his house—at night (even though, of course, darkness made no difference to him).

And he felt that his newly strengthened visual imagery enabled him to think in ways that had not been available to him before, allowed him to project himself inside machines and other systems, to envisage solutions, models, and designs.

I wrote back to Torey, suggesting that he consider writing another book, a more personal one, exploring how his life had been affected by blindness and how he had responded to this in the most improbable and seemingly paradoxical way. A few years later, he sent me the manuscript of **Out of Darkness.** In this new book, Torey described the early visual memories of his childhood and youth in Hungary before the Second World War: the sky-blue buses of

Budapest, the egg-yellow trams, the lighting of gas lamps, the funicular railway on the Buda side. He described a carefree and privileged youth, roaming with his father in the wooded mountains above the Danube, playing games and pranks at school, growing up in a highly intellectual environment of writers, actors, professionals of every sort. Torey's father was the head of a large motion-picture studio and would often give his son scripts to read. "This," Torey wrote, "gave me the opportunity to visualize stories, plots and characters, to work my imagination—a skill that was to become a lifeline and source of strength in the years ahead."

All of this came to a brutal end with the Nazi occupation, the siege of Buda, and then the Soviet occupation. Torey, by this time an adolescent, found himself passionately drawn to the big questions—the mystery of the universe, of life, and, above all, the mystery of consciousness, of the mind. At nineteen, feeling that he needed to immerse himself in biology, engineering, neuroscience, and psychology, but knowing that there was no chance of an intellectual life in Soviet Hungary, Torey made his escape and found his way to Australia, where, penniless and without connections, he did various manual jobs. In June of 1951, loosening the plug in a vat of acid at the chemical factory where he worked, he had the accident that bisected his life:

> The last thing I saw with complete clarity was a glint of light in the flood of acid that was to engulf my face and change my life. It was a nano-second of sparkle, framed by the black circle of the drumface, less than a foot away. This was the final scene, the slender thread that ties me to my visual past.

When it became clear that his corneas had been hopelessly damaged and that he would have to live his life as a blind man, he was advised to rebuild his representation of the world on the basis of hearing and touch, and to "forget about sight and visualizing altogether." But this was something that Torey could not or would not do. He had emphasized, in his first letter to me, the importance of a most critical choice at this juncture: "I immediately resolved to find out how far a partially sense-deprived brain could go to rebuild a life." Put this way, it sounds abstract, like an experiment. But in his book one senses the tremendous feelings underlying his resolution: the horror of darkness—"the empty darkness," as Torey often calls it, "the grey fog that was engulfing me"—and the passionate desire to hold on to light and sight, to maintain, if only in memory and imagination, a vivid and living visual world. The very title of his book says all this, and the note of defiance is sounded from the start.

Hull, who did not use his imagery in a deliberate way, lost it within two or three years and became unable to remember which way round a 3 went; Torey, on the other hand, soon became able to multiply four-figure numbers by each other, as on a blackboard, visualizing the whole operation in his mind, "painting" the sub-operations in different colors.

Torey maintained a cautious and "scientific" attitude to his own visual imagery, taking pains to check the accuracy of his images by every means available. "I learned," he wrote, "to hold the image in a tentative way, conferring credibility and status on it only when some information would tip the balance in its favor." He soon gained enough confidence in the reliability of his visual imagery to stake his life upon it, as when he undertook roof repairs by himself. And this confidence extended to other, purely mental projects. He became able "to imagine, to visualize, for example, the inside of a differential gearbox in action as if from inside its casing. I was able to watch the cogs bite, lock and revolve, distributing the spin as required. I began to play around with this internal view in connection with mechanical and technical problems, visualizing how subcomponents relate in the atom, or in the living cell." This power of imagery was crucial, Torey thought, in enabling him to arrive at a new view of the brain-mind problem by visualizing the brain "as a perpetual juggling act of interacting routines."

Soon after receiving the manuscript of **Out of Darkness,** I received proofs of yet another memoir about blindness: Sabriye Tenberken's **My Path Leads to Tibet.** While Hull and Torey are thinkers, preoccupied in their different ways by inwardness, states of brain and mind, Tenberken is a doer; she has traveled, often alone, all over Tibet, where for centuries blind people have been treated as less than human and denied education, work, respect, or a role in the community. Virtually single-handed, Tenberken has transformed their situation over the past decade or so, devising a form of Tibetan Braille, establishing the first schools for the blind there, and integrating the graduates of these schools into their communities.

Tenberken herself had impaired vision almost from birth, but was able to make out faces and landscapes until she was twelve. As a child in Germany, she loved painting and had a particular predilection for colors, and when she was no longer able to decipher shapes and forms, she could still use colors to identify objects.[3]

Though she had been totally blind for a dozen years when she went to Tibet, Tenberken continued to use her other senses, along with verbal descriptions, visual memories, and a strong pictorial and synesthetic sensibility, to construct "pictures" of landscapes and rooms, of environments and scenes—pictures so lively and detailed as to astonish her listeners. These images may sometimes be wildly or comically different from reality, as she related in one incident when she and a companion drove to Nam Co, the great salt lake in Tibet. Turning eagerly towards the lake, Tenberken saw, in her imagination, "a beach of crystallized salt shimmering like snow under an evening sun, at the edge of a vast body of turquoise water. . . . And down below, on the deep green mountain flanks, a

---

3. Tenberken also has an intense synesthesia, which has persisted and been intensified, it seems, by her blindness:

> As far back as I can remember, numbers and words have instantly triggered colors in me. . . . The number 4, for example, [is] gold. Five is light green. Nine is vermilion. . . . Days of the week as well as months have their colors, too. I have them arranged in geometrical formations, in circular sectors, a little like a pie. When I need to recall on which day a particular event happened, the first thing that pops up on my inner screen is the day's color, then its position in the pie.

few nomads were watching their yaks grazing." It then turned out that she had not been "looking" at the lake at all, but facing in another direction, "staring" at rocks and a gray landscape. These disparities do not faze her in the least—she is happy to have so vivid a visual imagination. Hers is essentially an artistic imagination, which can be impressionistic, romantic, not veridical at all, whereas Torey's imagination is that of an engineer, and has to be factual, accurate down to the last detail.

Jacques Lusseyran was a French Resistance fighter whose memoir, **And There Was Light**, deals mostly with his experiences fighting the Nazis and later in Buchenwald, but includes many beautiful descriptions of his early adaptations to blindness. He was blinded in an accident when he was not quite eight years old, an age that he came to feel was "ideal" for such an eventuality, for, while he already had a rich visual experience to call on, "the habits of a boy of eight are not yet formed, either in body or in mind. His body is infinitely supple."

At first, Lusseyran began to lose his visual imagery:

> A very short time after I went blind I forgot the faces of my mother and father and

> the faces of most of the people I loved. . . . I stopped caring whether people were dark or fair, with blue eyes or green. I felt that sighted people spent too much time observing these empty things. . . . I no longer even thought about them. People no longer seemed to possess them. Sometimes in my mind men and women appeared without heads or fingers.

This is similar to Hull, who wrote, "Increasingly, I am no longer even trying to imagine what people look like. . . . I am finding it more and more difficult to realize that people look like anything, to put any meaning into the idea that they have an appearance."

But then, while relinquishing the actual visual world and many of its values and categories, Lusseyran began to construct and to use an imaginary visual world more like Torey's. He came to identify himself as belonging to a special category, the "visual blind."

Lusseyran's inner vision started as a sensation of light, a formless, flooding, streaming radiance. Neurological terms are bound to sound reductive in this almost mystical context, yet one might venture to interpret this as a release phenomenon, a spontaneous, almost eruptive arousal of the visual cortex, now deprived of its normal visual input. (Such a phenomenon is analogous, perhaps, to tinnitus or phantom limbs, though endowed, here, by a devout and precociously imaginative little boy, with some element of the supernal.) But then, it becomes clear, he found himself in possession of great powers of visual imagery, and not just a formless luminosity.

The visual cortex, the inner eye, having been activated, his mind constructed a "screen" upon which whatever he thought or desired was projected and, if need be, manipulated, as on a computer screen. "This screen was not like a blackboard, rectangular or square, which so quickly reaches the edge of its frame," he wrote.

> My screen was always as big as I needed it to be. Because it was nowhere in space it was everywhere at the same time. . . . Names, figures and objects in general did not appear on my screen without shape, nor just in black and white, but in all the colors of the rainbow. Nothing entered my mind without being bathed in a certain amount of light. . . . In a few months my personal world had turned into a painter's studio.

Great powers of visualization were crucial to the young Lusseyran, even in something as nonvisual (one would think) as learning Braille, and in his brilliant successes at school.

Visualization was no less crucial in the real, outside world. Lusseyran described walks with his sighted friend Jean, and how, as they were climbing together up the side of a hill above the Seine Valley, he could say to Jean:

> "Just look! This time we're on top. . . . You'll see the whole bend of the river, unless the sun gets in your eyes!" Jean was startled, opened his eyes wide and cried: "You're right." This little scene was often repeated between us, in a thousand forms.
>
> Every time someone mentioned an event, the event immediately projected itself in its place on the screen, which was a kind of inner canvas. . . . Comparing my world with his, [Jean] found that his held fewer pictures and not nearly as many colors. This made him almost angry. "When it comes to that," he used to say, "which one of us two is blind?"

It was his supernormal powers of visualization and visual manipulation—visualizing people's positions and movements, the topography of any space, visualizing strategies for defense and attack—coupled with his charismatic personality (and seemingly infallible "nose" or "ear" for detecting possible traitors) that later made Lusseyran an icon in the French Resistance.

I had now read four memoirs, all strikingly

different in their depictions of the visual experience of blinded people: Hull with his acquiescent descent into "deep blindness"; Torey with his "compulsive visualization" and meticulous construction of an internal visual world; Tenberken with her impulsive, almost novelistic visual freedom, along with her remarkable and specific gift of synesthesia; and Lusseyran, who identified himself as one of the "visual blind." Was there any such thing, I wondered, as a typical blind experience?

Dennis Shulman, a clinical psychologist and psychoanalyst who lectures on biblical topics, is an affable, stocky, bearded man in his fifties who gradually lost his sight in his teens, becoming completely blind by the time he entered college. When we met a few years ago, he told me that his experience was completely unlike Hull's:

> I still live in a visual world after thirty-five years of blindness. I have very vivid visual memories and images. My wife, whom I have never seen—I think of her visually. My kids, too. I see myself visually—but it is as I last saw myself, when I was thirteen, though I try hard to update the image. I often give pub-

> lic lectures, and my notes are in Braille; but when I go over them in my mind, I see the Braille notes visually—they are visual images, not tactile.

Arlene Gordon, a former social worker in her seventies, told me that things were very similar for her. She said, "I was stunned when I read [Hull's book]. His experiences are so unlike mine." Like Dennis, she still identifies herself in many ways as a visual person. "I have a very strong sense of color," she said. "I pick out my own clothes. I think, 'Oh, that will go with this or that,' once I have been told the colors." Indeed, she was dressed very smartly, and took obvious pride in her appearance.

She still had a great deal of visual imagery, she continued: "If I move my arms back and forth in front of my eyes, I see them, even though I have been blind for more than thirty years." It seemed that moving her arms was immediately translated into a visual image. Listening to talking books, she added, made her eyes ache if she listened too long; she felt herself to be "reading" at such times, the sound of the spoken words being transformed to lines of print on a vividly visualized book in front of her.[4]

---

4. Although I myself am a poor visualizer, if I shut my eyes, I can still "see" my hands moving on the piano keyboard when I

Arlene's comment reminded me of Amy, a patient who had been deafened by scarlet fever at the age of nine but was so adept a lip-reader that I often forgot she was deaf. Once, when I absentmindedly turned away from her as I was speaking, she said sharply, "I can no longer hear you."

"You mean you can no longer see me," I said.

"**You** may call it seeing," she answered, "but I experience it as hearing."

Amy, though totally deaf, still constructed the sound of speech in her mind. Both Dennis and Arlene, similarly, spoke not only of a heightening of visual imagery and imagination since losing their eyesight but also of what seemed to be a much readier transference of information from verbal description—or from their own sense of touch, movement, hearing, or smell—into a visual form. On the whole, their experiences seemed quite similar to Torey's, even though they had not systematically exercised

---

play a piece that I know well. (This may happen even if I just play the piece in my mind.) I feel my hands moving at the same time, and I am not entirely sure that I can distinguish the "feeling" from the "seeing." In this context, they seem inseparable, and one wants to use an intersensory term like "seeing-feeling."

The psychologist Jerome Bruner speaks of such imagery as "enactive"—an integral feature of a performance (real or imaginary)—in contrast to an "iconic" visualization, the visualization of something outside oneself. The brain mechanisms underlying these two sorts of imagery are quite different.

their powers of visual imagery the way he had, or consciously tried to make an entire virtual world of sight.

What happens when the visual cortex is no longer limited or constrained by any visual input? The simple answer is that, isolated from the outside, the visual cortex becomes hypersensitive to internal stimuli of all sorts: its own autonomous activity; signals from other brain areas—auditory, tactile, and verbal areas; and thoughts, memories, and emotions.

Torey, unlike Hull, played a very active role in building up his visual imagery, took control of it the moment the bandages were removed. Perhaps this was because he was already very at home with visual imagery, and used to manipulating it in his own way. We know that Torey was very visually inclined before his accident, and skilled from boyhood in creating visual narratives based on the film scripts his father gave him. (We have no such information about Hull, for his journal entries start only when he has become blind.)

Torey required months of intense cognitive discipline dedicated to improving his visual imagery, making it more tenacious, more stable, more malleable, whereas Lusseyran seemed to do this almost from the start. Perhaps this was because Lusseyran was not yet eight when blinded (while Torey was twenty-one), and his brain was, accordingly, more able to adapt to a new and drastic contingency. But adaptability does not end with youth. It is clear that Arlene, who became blind in her forties, was able to adapt in quite radical ways, too, developing the ability to "see" her hands moving before her, to "see" the words of books read to her, to construct detailed visual images from verbal descriptions. One has a sense that Torey's adaptation was largely shaped by conscious motive, will, and purpose; that Lusseyran's was shaped by overwhelming physiological disposition; and that Arlene's lies somewhere in between. Hull's, meanwhile, remains enigmatic.

How much do these differences reflect an underlying predisposition independent of blindness? Do sighted people who are good visualizers, who have strong visual imagery, maintain or even enhance their powers of imagery if they become blind? Do people who are poor visualizers, on the other hand, tend to move towards "deep blindness" or hallucinations if they lose their sight? What is the range of visual imagery in the sighted?

I first became conscious of great variations in the power of visual imagery and visual memory when I was fourteen or so. My mother was a surgeon and comparative anatomist, and I had brought her a lizard's skeleton from school. She gazed at this intently for a minute, turning it round in her hands, then put it down and without looking at it again did a number of drawings of it, rotating it mentally by thirty degrees each time, so that she produced a series, the last drawing exactly the same as the first. I could not imagine how she had done this. When she said that she could see the skeleton in her mind just as clearly and vividly as if she were looking at it, and that she simply rotated the image through a twelfth of a circle each time, I felt bewildered, and very stupid. I could hardly see anything with my mind's eye—at most, faint, evanescent images over which I had no control.[5]

---

5. Though I have almost no voluntary imagery, I am prone to involuntary imagery. I used to have this only as I was falling asleep, in migraine auras, with some drugs, or with fever. But now that my sight is impaired, I have it all the time.

In the 1960s, during a period of experimenting with large doses of amphetamines, I experienced a different sort of vivid mental imagery. Amphetamines can produce striking perceptual changes and dramatic enhancements of visual imagery and memory (as I described in "The Dog Beneath the Skin," a chapter in **The Man Who Mistook His Wife for a Hat**). For a period of two weeks or so, I found that I had only to look

My mother had hoped I would follow in her footsteps and become a surgeon, but when she realized how lacking in visual powers I was (and how clumsy, lacking in mechanical skill, too) she resigned herself to the idea that I would have to specialize in something else.

A few years ago, at a medical conference in Boston, I spoke about Torey's and Hull's experiences of blindness, how "enabled" Torey seemed to be by the powers of visualization he had developed, and how "disabled" Hull was—in some ways, at least—by the loss of his powers of visual imagery and memory. After my talk, a man in the audience came up to me and asked how well, in my estimation, sighted people could function if they had no visual imagery. He went on to say that he had no visual

---

at an anatomical picture or specimen, and its image would remain vivid and stable in my mind for hours. I could mentally project the image onto a piece of paper—it was as clear and distinct as if projected by a camera lucida—and trace its outlines with a pencil. My drawings were not elegant, but they were, everyone agreed, quite detailed and accurate. But when the amphetamine-induced state faded, I could no longer visualize, no longer project images, no longer draw—nor have I been able to do so in the decades since. This was not like voluntary imagery—I did not summon images to my mind or construct them bit by bit. It was involuntary and automatic, more akin to eidetic or "photographic" memory, or to palinopsia, an exaggerated persistence of vision.

imagery whatever, at least none that he could deliberately evoke, and that no one in his family had any, either. Indeed, he had assumed this was the case with everyone until, as a student at Harvard, he had come to participate in some psychological tests and had realized that he apparently lacked a mental power that all the other students, in varying degrees, had.

"And what do you do?" I asked him, wondering what this poor man **could** do.

"I am a surgeon," he replied. "A vascular surgeon. An anatomist, too. And I design solar panels." But how, I asked him, did he recognize what he was seeing?

"It's not a problem," he answered. "I guess there must be representations or models in the brain that get matched up with what I am seeing and doing. But they are not conscious. I cannot evoke them."

This seemed to be at odds with my mother's experience—she, clearly, did have extremely vivid and readily manipulable visual imagery, though (it now seemed) this may have been a bonus, a luxury, and not a prerequisite for her career as a surgeon.

Is this also the case with Torey? Is his greatly developed visual imagery, though clearly a source of much pleasure, not as indispensable as he takes it to be? Might he, in fact, have been able to do everything he did, from carpentry to roof repair to making a model of the mind, without any conscious imagery at all? He himself raises this question.

The role of mental imagery in thinking was explored by Francis Galton in his 1883 book **Inquiries into Human Faculty and Its Development.** (Galton, a cousin of Darwin's, was irrepressible and wide-ranging, and his book includes chapters on subjects as various as fingerprints, eugenics, dog whistles, criminality, twins, synesthesia, psychometric measures, and hereditary genius.) His inquiry into voluntary visual imagery took the form of a questionnaire, with such questions as "Can you recall with distinctness the features of all near relations and many other persons? Can you at will cause your mental image . . . to sit, stand, or turn slowly around? Can you . . . see it with enough distinctness to enable you to sketch it leisurely (supposing yourself able to draw)?" The vascular surgeon would have been hopeless on such tests—indeed, it was questions such as these that had floored him when he was a student at Harvard. And yet, finally, how much had it mattered?

As to the significance of such imagery, Galton is ambiguous and guarded. He suggests, in one breath, that "scientific men, as a class, have feeble powers of visual representation" and, in another, that "a vivid visualizing faculty is of much importance in connection with the higher processes of generalized thoughts." He feels that "it is undoubtedly the fact that mechanicians, engineers and architects usually possess the faculty of seeing mental images with remarkable clearness and precision" but adds, "I am, however, bound to say, that the missing faculty seems to be replaced so serviceably by other modes of conception . . . that men who declare themselves entirely deficient in the power of seeing mental pictures can nevertheless give lifelike descriptions of what they have seen, and can otherwise express themselves as if they were gifted with a vivid visual imagination. They can also become painters of the rank of Royal Academicians."

A mental image, for Galton, was picturing a familiar person or place in the mind's eye; it was a reproduction or reconstruction of an experience. But there are also mental images of a much more abstract and visionary kind, images of something which has never been seen by the physical eye but which can be conjured up by the creative imagination and serve as models for investigating reality.[6]

---

6. The physicist John Tyndall referred to these in an 1870 lecture, a few years before Galton's **Inquiries:** "In explain-

In his book **Image and Reality: Kekulé, Kopp, and the Scientific Imagination**, Alan Rocke brings out the crucial role of such images or models in the creative lives of scientists, especially nineteenth-century chemists. He focuses especially on August Kekulé and the famous reverie, while he was riding a London bus, that led him to visualize the structure of a benzene molecule, a concept that would revolutionize chemistry. Although chemical bonds are invisible, they were as real to Kekulé, as visually imaginable, as the lines of force around a magnet were for Faraday. Kekulé said of himself that he had "an irresistible need for visualization."

Indeed, a conversation about chemistry can hardly be maintained without such images and models, and in **Mindsight**, the philosopher Colin McGinn writes, "Images are not just minor variations on perception and thought, of negligible theoretical interest; they are a robust mental category in need of independent investigation. . . . Mental images . . . should be added as a third great category . . . to the twin pillars of perception and cognition."

---

ing scientific phenomena, we habitually form mental images of the ultra-sensible. . . . Without the exercise of this power our knowledge of nature would be a mere fabulation of coexistences and sequences."

Some people, like Kekulé, are clearly very powerful visualizers in this abstract sense, but most of us use some combination of experiential visualization (imaging one's house, for example) and abstract visualization (imagining the structure of an atom). Temple Grandin, though, feels she is a different sort of visualizer.[7] She thinks entirely in terms of literal images she has seen before, as if she is looking at a familiar photograph or a film running in her head. When she imagines the concept of "heaven," for instance, her instant association is to the film **Stairway to Heaven**, and the image in her mind is that of a staircase ascending into the clouds. If someone remarks that it is a rainy day, she sees, in her mind's eye, the same "photograph" of rain, her own literal and iconic representation of rain. Like Torrey, she is a powerful visualizer; her extremely accurate visual memory allows her to walk through, in her mind, a factory she is designing, noting structural details even before it is built. Growing up, she assumed this was how everyone thought, and she is puzzled, now, by the idea that some people cannot

---

7. I described Temple more fully in **An Anthropologist on Mars**, and she speaks about her visual thinking especially in her book **Thinking in Pictures**.

summon visual images at will. When I told her I could not do so, she asked, "How **do** you think, then?"

When I talk to people, blind or sighted, or when I try to think of my own internal representations, I find myself uncertain whether words, symbols, and images of various types are the primary tools of thought or whether there are forms of thought antecedent to all of these, forms of thought essentially amodal. Psychologists have sometimes spoken of "interlingua" or "mentalese," which they conceive to be the brain's own language, and Lev Vygotsky, the great Russian psychologist, used to speak of "thinking in pure meanings." I cannot decide whether this is nonsense or profound truth—it is the sort of reef I end up on when I think about thinking.

Galton himself was puzzled about visual imagery: it had an enormous range, and although it sometimes seemed an essential part of thinking, at other times it seemed irrelevant. This uncertainty has characterized the debate over mental imagery ever since. A contemporary of Galton's, the early experimental psychologist Wilhelm Wundt, guided by introspection, believed imagery to be an essential part of thought. Others maintained that thinking was imageless and consisted entirely of analytical or descriptive propositions, and behaviorists did not believe in thinking at all—there was only "behavior." Was introspection alone a reliable method of scientific observation? Could it yield data that were consistent, repeatable, measurable? It was only in the early 1970s that this challenge was faced by a new generation of psychologists. Roger Shepard and Jacqueline Metzler asked subjects to perform mental tasks that required rotating an image of a geometrical figure in their minds—the sort of imaginary rotation my mother performed when she drew the lizard's skeleton from memory. They were able to determine in these first quantitative experiments that rotating an image took a specific amount of time—an amount proportional to the degree of rotation. Rotating an image through sixty degrees, for instance, took twice as long as rotating it through thirty degrees, and rotating it through ninety degrees, three times as long. Mental rotation had a rate, it was continuous and steady, and it took effort, like any voluntary act.

Stephen Kosslyn entered the subject of visual imagery from another angle, and in 1973 published a seminal paper contrasting the performance of "imagers" and "verbalizers" who were asked to remember a set of drawings they had been shown. Kosslyn hypothesized that if internal images were spatial and organized like pictures, the "imagers" ought to be able to focus selectively on a part of the image, and that time would be required for them to shift their attention from one part of the image to another. The time required, he thought, would be proportional to the distance the mind's eye had to travel.

Kosslyn was able to show that all of these were indeed the case, indicating that visual images were essentially spatial and organized in space like pictures. His work has proved immensely fertile, but the ongoing debate about the role of visual imagery continues, as Zenon Pylyshyn and others have maintained that the mental rotation of images and "scanning" them could be interpreted as the result of purely abstract, nonvisual operations in the mind/brain.[8]

By the 1990s, Kosslyn and others were able to combine imagery experiments with PET and fMRI scanning, which allowed them to map the areas of the brain involved as people engaged in tasks requiring mental imagery. Mental imagery, they found, activated many of the same areas of the visual cortex as percep-

---

8. Kosslyn's latest book on the matter, **The Case for Mental Imagery**, details the history of this debate.

tion itself, showing that visual imagery was a physiological reality as well as a psychological one, and used at least some of the same neural pathways as visual perception.[9]

That perception and imagery share a common neural basis in the visual parts of the brain is suggested by clinical studies, too. In 1978 Eduardo Bisiach and Claudio Luzzatti in Italy related the cases of two patients who both developed a hemianopia following a stroke and could not see to the left side. When they were asked to imagine themselves walking down a familiar street and describe what they saw, they mentioned only the shops on the right side of the street; but when they were then asked to imagine turning around and walking back, they described the shops they had not "seen" before, the shops that were now on the right side. These beautifully examined cases showed that a hemianopia might cause not only a bi-

9. Functional MRIs also showed that the two hemispheres of the brain behaved differently in regard to imagery, the left hemisphere concerned with generic, categorical images—e.g., "trees"—and the right hemisphere with specific images—e.g., "the maple in my front yard"—a specialization also present in visual perception. Thus prosopagnosia, an inability to recognize specific faces, is associated with damaged or defective visual function in the right hemisphere, though people with prosopagnosia have no problem with the category of faces in general, a left-hemisphere function.

section of the visual field but a bisection of visual imagery as well.

Such clinical observations on the parallels between visual perception and visual imagery go back at least a century. In 1911, the English neurologists Henry Head and Gordon Holmes examined a number of patients with subtle damage to the occipital lobes—damage that led not to total blindness but to blind spots within the visual field. They found, by questioning their patients carefully, that blind spots in exactly the same locations occurred in the patients' mental imagery as well. And in 1992, Martha Farah et al. reported that in a patient who lost partial vision on one side due to an occipital lobectomy, the visual angle of his mind's eye was also reduced, in a way that perfectly matched his perceptual loss.

For me, the most convincing demonstration that at least some aspects of visual imagery and visual perception might be inseparable occurred when I was consulted in 1986 by Mr. I., an artist who became completely colorblind following a head injury.[10] Mr. I. was distressed by his sudden inability to perceive colors, but even more by his total inability to evoke them in memory or imagery. Even his occasional

---

10. Mr. I.'s case is described in **An Anthropologist on Mars.**

visual migraines were now drained of color. Patients like Mr. I. suggest that the coupling of perception and imagery is very close in the higher parts of the visual cortex.[11]

Sharing characteristics and even sharing neural areas or mechanisms is one thing, but Kosslyn and others go further than this, sug-

---

11. While it seems clear that perception and imagery share certain neural mechanisms at higher levels, this sharing is less evident in the primary visual cortex—hence the possibility of a dissociation such as occurs in Anton's syndrome. In Anton's syndrome, patients with occipital damage are cortically blind, but believe they are still sighted. They will move about without restraint or caution, and if they bump into a piece of furniture, they will ascribe this, perhaps, to the furniture being "out of place."

Anton's syndrome is sometimes attributed to the preservation of some visual imagery despite occipital damage, and to patients mistaking this imagery for perception. But there may be other, stranger mechanisms at work. The denial of blindness—or, more accurately, the inability to realize that one has lost one's vision—is very like another "disconnection syndrome," known as anosognosia. With anosognosia, following damage to the right parietal lobe, patients lose awareness of their left side, and of the left half of space, along with the awareness that anything is amiss. If one draws their attention to their left arm, they will say it is someone else's—"the doctor's arm," or "my brother's arm," or even "an arm someone left here." Such confabulations seem similar in a way to those of Anton's syndrome, attempts to explain what, to the patient, is a bizarrely inexplicable situation.

gesting that visual perception **depends** on visual imagery, matching what the eye sees, the retina's output, with memory images in the brain. Visual recognition, they feel, could not occur without such matching. Kosslyn proposes, furthermore, that mental imagery may be crucial in thought itself—problem solving, planning, designing, theorizing. Support for this comes from studies asking subjects to answer questions that would seem to require visual imagery—for example, "Which is darker green, a frozen pea or a pine tree?" or "What shape are Mickey Mouse's ears?" or "In which hand does the Statue of Liberty hold her torch?"—or asking them to solve problems that can be worked out either by means of imagery or by means of more abstract, nonvisual thinking. Kosslyn speaks here of a doubleness in the way people think, contrasting the use of "depictive" representations, which are direct and unmediated, with "descriptive" ones, which are analytic and mediated by verbal or other symbols. Sometimes, he suggests, one mode will be favored over another, depending on the individual and on the problem to be solved. Sometimes both modes will proceed in tandem (although depiction is likely to outpace description), and at other times one may start with depiction—images—and proceed to a purely verbal or mathematical representation.[12]

What, then, of people like me, or the vascular surgeon in Boston who cannot evoke **any** visual images voluntarily? One must infer, as my colleague in Boston does, that we, too, have visual images, models, and representations in the brain, images that allow visual perception and recognition but are below the threshold of consciousness.[13]

---

12. Einstein described this in regard to his own thinking:

> The psychical entities which seem to serve as elements in thought are certain signs and more or less clear images which can be "voluntarily" reproduced and combined. . . . [Some] are, in my case, of visual and some of muscular type. Conventional words or other signs have to be sought for laboriously only in a second stage.

Darwin, on the other hand, seemed to describe a very abstract, almost computational process in his own thinking, when he wrote in his autobiography, "My mind seems to have become a kind of machine for grinding general laws out of large collections of facts." (What Darwin omitted here was that he had a fantastic eye for form and detail, an enormous observational and depictive power, and it was these which provided the "facts.")

13. Dominic ffytche, who has investigated the neurobiology of conscious vision—imagery and hallucination as well as perception—feels that visual consciousness is a threshold phenomenon. Using fMRIs to study patients with visual hallucinations, he has shown that there may be evidence of unusual activity in a specific part of the visual system—for example, the fusiform

· · ·

If the central role of visual imagery is to permit visual perception and recognition, what need is there for it if a person becomes blind? And what happens to its neural substrates, the visual areas which occupy nearly half of the entire cerebral cortex? We know that in adults who lose their eyesight, there may be some atrophy of the pathways and relay centers leading from the retina to the cerebral cortex—but there is little degeneration in the visual cortex itself. Functional MRIs of the visual cortex show no diminution of activity in such a situation; indeed, we see the reverse: they reveal a heightened activity and sensitivity. The visual cortex, deprived of visual input, is still good neural real estate, available and clamoring for a new function. In someone like Torey, this may free up more cortical space for visual imagery; in someone like Hull, relatively more may be employed by other senses—auditory perception and attention, perhaps, or tactile perception and attention.[14]

---

face area—but this has to reach a certain intensity before it enters consciousness, before the subject actually "sees" faces.

14. The heightened (and sometimes morbid) sensitivity of the visual cortex when deprived of its normal perceptual input

This sort of cross-modal activation may underlie the fact that some blind people, like Dennis Shulman, "see" Braille as they read it with their finger. This may be more than just an illusion or a fanciful metaphor; it may be a reflection of what is actually happening in his brain, for there is good evidence that reading Braille can cause strong activation of the visual parts of the cortex, as Sadato, Pascual-Leone,

---

may also predispose it to intrusive imagery. A significant proportion of those who go blind—10 to 20 percent, by most estimates—become prone to involuntary images, or outright hallucinations, of an intense and sometimes bizarre kind. Such hallucinations were originally described in the 1760s by the Swiss naturalist Charles Bonnet, and we now speak of hallucinations secondary to visual impairment as Charles Bonnet syndrome.

Hull described something akin to this which occurred for a while after he lost the last of his sight:

> About a year after I was registered blind, I began to have such strong images of what people's faces looked like that they were almost like hallucinations. . . . I would be sitting in a room with someone, my face pointed towards my companion, listening to him or her. Suddenly, such a vivid picture would flash before my mind that it was like looking at a television set. Ah, I would think, there he is, with his glasses and his little beard, his wavy hair and his blue, pinstriped suit, white collar and blue tie. . . . Now this image would fade and in its place another one would be projected. My companion was now fat and perspiring with receding hair. He had a red necktie and waistcoat, and a couple of his teeth were missing.

et al. have reported. Such activation, even in the absence of any input from the retina, may constitute a crucial part of the neural basis for the mind's eye.

Dennis also spoke of how the heightening of his other senses had increased his sensitivity to the most delicate nuances in other people's speech and self-presentation. He could recognize many of his patients by smell, he said, and he could often pick up states of tension or anxiety they might not even be aware of. He felt that he had become far more sensitive to others' emotional states since losing his sight, for he was no longer taken in by visual appearances, which most people learn to camouflage. Voices and smells, by contrast, he felt, could reveal people's depths.

The heightening of other senses with blindness allows a number of very remarkable adaptations, including "facial vision," the ability to use sound or tactile clues to sense the shape or size of a space and the people or objects in it.

Martin Milligan, the philosopher, who had both eyes removed at the age of two (because of malignant tumors), has written of his own experience:

> Born-blind people with normal hearing don't just hear sounds: they can hear objects (that

> is, have an awareness of them, chiefly through their ears) when they are fairly close at hand, provided these objects are not too low; and they can also in the same way "hear" something of the shape of their immediate surroundings. . . . Silent objects such as lamp-posts and parked cars with their engines off can be heard by me as I approach them and pass them as atmosphere-thickening occupants of space, almost certainly because of the way they absorb and/or echo back the sounds of my footsteps and other small sounds. . . . It isn't usually necessary to make sounds oneself to have this awareness, though it helps. Objects of head height probably slightly affect the air currents reaching my face, which helps towards my awareness of them—which is why some blind people refer to this kind of sense-awareness as their "facial" sense.

Facial vision tends to be most highly developed in those who are born blind or lose their sight at an early age; for the writer Ved Mehta, who has been blind since the age of four, it is so well developed that he walks confidently and rapidly without a cane, and it is sometimes difficult for others to realize that he is blind.

While the sound of one's footsteps or one's cane may suffice, other forms of echolocation have been reported. Ben Underwood developed an astonishing, dolphinlike strategy of emitting regular clicks with his mouth and accurately reading the resulting echoes from nearby objects. He was so adept at moving about the world in this way that he was able to play field sports and even chess.[15]

Blind people often say that using a cane enables them to "see" their surroundings, as touch, action, and sound are immediately transformed into a "visual" picture. The cane acts as a sensory substitution or extension. But is it possible to give a blind person a more detailed picture of the world, using more modern technology? Paul Bach-y-Rita was a pioneer in this realm and spent decades testing all sorts of sensory substitutes, though his special interest lay in developing devices that could help the blind by using tactile images. (In 1972, he published a prescient book surveying all the possible brain mechanisms by which sensory substitution might be realized. Such substitution, he emphasized, would depend on the brain's plasticity—and that the brain had any plasticity at all was a revolutionary concept at the time.)

---

15. Ben, who had retinoblastoma, had both eyes removed at the age of three, but then, tragically, died at sixteen from a recurrence of his cancer. Videos of Ben and his echolocation can be seen at the website www.benunderwood.com.

Bach-y-Rita wondered if one might connect the output of a video camera, point by point, to the skin, allowing a blind subject to form a "touch picture" of his environment. This might work, he thought, because tactile information is organized topographically in the brain, and topographic accuracy is essential for forming a quasi-visual picture. Eventually, he began using tiny grids of a hundred or so electrodes on that most sensitive part of the body, the tongue. (The tongue has the highest density of sensory receptors in the body, and it also occupies the greatest amount of space, proportionally, in the sensory cortex. This makes it uniquely suitable for sensory substitution.) With this device, the size of a postage stamp, his subjects could form a crude but nevertheless useful "picture" on the tongue itself.

Over the years, the sophistication of such devices has increased greatly, and prototypes now have four to six times the resolution of Bach-y-Rita's early version. Bulky camera cables have been replaced by spectacles containing miniature cameras, allowing subjects to direct the cameras by a more natural head movement. With this, blind subjects are able to walk across a room that is not too cluttered, or to catch a ball rolled towards them.

Does this mean that they are now "seeing"?

Certainly, they are showing what behaviorists would call "visual behavior." Bach-y-Rita spoke of how his subjects "learn[ed] to make perceptual judgements using visual means of interpretation, such as perspective, parallax, looming and zooming, and depth estimates." Many of these people **felt** as if they were seeing once again, and functional MRIs showed strong activations of visual areas in their brains while they were "seeing" with the camera. ("Seeing" occurred particularly when the subjects were able to move the camera voluntarily, pointing it here or there, **looking** with it. Looking was crucial, for there is no perception without action, no seeing without looking.)

To restore sight to someone who once had it, whether by surgical means or by a sensory-substitution device, is one thing, for such a person would have an intact visual cortex and a lifetime of visual memories. But to give sight to someone who has never seen, never experienced light or sight, would seem to be impossible, in view of what we know about the brain's critical periods and the necessity of at least some visual experience in the first two years of life to stimulate the development of the visual cortex. (Recent work from Pawan Sinha and others, however, suggests that the critical period may not be as critical as previously accepted.)[16] Tongue vision has been tried with congenitally blind people, too, and with some success. One young musician, born blind, said she "saw" the conductor's gestures for the first time in her life.[17] Although the visual cortex in congenitally blind people is reduced in volume by more than 25 percent, it can still, apparently, be activated by sensory substitution, and this has been confirmed, in several cases, by fMRIs.[18]

---

16. See Ostrovsky et al., for example.

17. Congenitally blind people, we might suppose, can have no visual imagery at all, since they have never had any visual experience. And yet they sometimes report having clear and recognizable visual elements in their dreams. Helder Bértolo and his colleagues in Lisbon, in an intriguing 2003 report, described how they compared congenitally blind subjects with normal sighted subjects and found "equivalent visual activity" (based on analysis of EEG alpha-wave attenuation) in the two groups while dreaming. The blind subjects were able, upon waking, to draw the visual components of their dreams, although they had a lower rate of dream recall. Bértolo et al. conclude, therefore, that "the congenitally blind have visual content in their dreams."

18. Would acquiring "sight" if one has never seen before be bewildering or enriching? For my patient Virgil, who was given sight, through surgery, after a lifetime of blindness, it was utterly incomprehensible at first, as I described in **An Anthropologist on Mars.** Thus although sensory-substitution technologies are exciting and promise a new freedom for blind people, we need

There is increasing evidence for the extraordinarily rich interconnectedness and interactions of the sensory areas of the brain, and the difficulty, therefore, of saying that anything is purely visual or purely auditory, or purely anything. The world of the blind can be especially rich in such in-between states—the intersensory, the metamodal—states for which we have no common language.[19]

---

to consider equally their impact on a life that has already been constructed without sight.

19. In a recent letter to his colleague Simon Hayhoe, John Hull expanded on this:

> For example, when the thought of a car occurs to me, although my front-line images are of recently touching the warm bonnet of a car, or of the shape of the car as I feel for the door handle, there are also traces of the appearance of the whole car, from pictures of cars in books, or memories of cars coming and going. Sometimes, when I have to touch a modern car, I am surprised to find that this memory trace does not correspond to reality, and that cars are not the same shape they were twenty-five years ago.
>
> There is a second point. The fact that an item of knowledge is so much buried in the sense or senses that first received it, means for me that I am not always sure whether my image is visual or not. The trouble is that tactile images of the shape and feel of things also often seem to acquire a visual content, or one cannot tell if the three-dimensional memory shape is being mentally represented by a visual or a tactile image. So even after all these years, the brain can't sort out where it is getting stuff from.

On Blindness is an exchange of letters between the blind philosopher Martin Milligan and a sighted philosopher, Bryan Magee. While his own nonvisual world seems coherent and complete to him, Milligan realizes that sighted people have access to a sense, a mode of knowledge, denied him. But congenitally blind people, he insists, can (and usually do) have rich and varied perceptual experiences, mediated by language and by imagery of a nonvisual sort. Thus they may have a "mind's ear" or a "mind's nose." But do they have a mind's eye?

Here Milligan and Magee cannot reach agreement. Magee insists that Milligan, a blind man, cannot have any real knowledge of the visual world. Milligan disagrees and maintains that even though language only describes people and events, it can sometimes stand in for direct experience or acquaintance.

Congenitally blind children, it has often been noted, tend to have superior memories and be precocious verbally. They may develop such fluency in the verbal description of faces and places as to leave others (and perhaps themselves) uncertain as to whether they are actually blind. Helen Keller's writing, to give a famous example, startles one with its brilliantly visual quality.

I loved reading Prescott's **Conquest of Mex-**

ico and **Conquest of Peru** as a boy, and felt that I "saw" these lands through his intensely visual, almost hallucinogenic descriptions. I was amazed to discover, years later, that Prescott had not only never visited Mexico or Peru; he had been virtually blind since the age of eighteen. Did he, like Torey, compensate for his blindness by developing huge powers of visual imagery, or were his brilliant visual descriptions simulated, in a way, made possible by the evocative and pictorial powers of language? To what extent can description, picturing in words, provide a substitute for actual seeing or for the visual, pictorial imagination?

After becoming blind in her forties, Arlene Gordon found language and description increasingly important; it stimulated her powers of visual imagery as never before and, in a sense, enabled her to see. "I love traveling," she told me. "I **saw** Venice when I was there." She explained how her traveling companions would describe places, and she would then construct a visual image from these details, her reading, and her own visual memories. "Sighted people enjoy traveling with me," she said. "I ask them questions, then they look and see things they wouldn't otherwise. Too often people with sight don't see anything! It's a reciprocal process—we enrich each other's worlds."

There is a paradox here—a delicious one—which I cannot resolve: if there is indeed a fundamental difference between experience and description, between direct and mediated knowledge of the world, how is it that language can be so powerful? Language, that most human invention, can enable what, in principle, should not be possible. It can allow all of us, even the congenitally blind, to see with another person's eyes.

Bibliography

Abbott, Edwin A. 1884. **Flatland: A Romance of Many Dimensions.** Reprint, New York: Dover, 1992.

Aguirre, Geoffrey K., and Mark D'Esposito. 1997. Environmental knowledge is subserved by separable dorsal/ventral neural areas. **Journal of Neuroscience** 17 (7): 2512–18.

Bach-y-Rita, Paul. 1972. **Brain Mechanisms in Sensory Substitution.** New York: Academic Press.

Bach-y-Rita, Paul, and Stephen W. Kercel. 2003. Sensory substitution and the human-machine interface. **Trends in Cognitive Sciences** 7 (12): 541–46.

Barry, Susan R. 2009. **Fixing My Gaze: A Scientist's Journey into Seeing in Three Dimensions.** New York: Basic Books.

Benson, D. Frank, R. Jeffrey Davis, and Bruce D. Snyder. 1988. Posterior cortical atrophy. **Archives of Neurology** 45 (7): 789–93.

Benson, D. Frank, and Norman Geschwind. 1969. The alexias. In **Handbook of Clinical Neurology,** vol. 4, ed. P. J. Vinken and G. W. Bruyn, pp. 112–40. Amsterdam: Elsevier.

Benton, Arthur L. 1964. Contributions to aphasia before Broca. **Cortex** 1: 314–27.

Berker, Ennis Ata, Ata Husnu Berker, and Aaron Smith. 1986. Translation of Broca's 1865 report: localization of speech in the third left frontal convolution. **Archives of Neurology** 43: 1065–72.

Bértolo, H. 2005. Visual imagery without visual perception? **Psicológica** 26: 173–88.

Bértolo, H., T. Paiva, L. Pessoa, T. Mestre, R. Marques, and R. Santos. 2003. Visual dream content, graphical representation and EEG alpha activity in congenitally blind subjects. **Brain Research/ Cognitive Brain Research** 15 (3): 277–84.

Beversdorf, David Q., and Kenneth M. Heilman. 1998. Progressive ventral posterior cortical degeneration presenting as alexia for music and words. **Neurology** 50: 657–59.

Bigley, G. Kim, and Frank R. Sharp. 1983. Reversible alexia without agraphia due to migraine. **Archives of Neurology** 40: 114–15.

Bisiach, E., and C. Luzzatti. 1978. Unilateral neglect of representational space. **Cortex** 14 (1): 129–33.

Bodamer, Joachim. 1947. Die Prosop-agnosie. **Archiv für Psychiatrie und Nervenkrankheiten** 179: 6–53.

Borges, Jorge Luis. 1984. Memories of a trip to Japan. In **Twenty-four Conversations with Borges,** ed. Roberto Alifano. Housatonic, MA: Lascaux Publishers.

Brady, Frank B. 2004. **A Singular View: The Art of Seeing with One Eye.** 6th ed. Vienna, VA: Michael O. Hughes.

Brewster, David. 1856. **The Stereoscope: Its History, Theory and Construction.** London: John Murray.

Campbell, Ruth. 1992. Face to face: interpreting a case of developmental prosopagnosia. In **Mental Lives: Case Studies in Cognition**, ed. Ruth Campbell, pp. 216–36. Oxford: Blackwell.

Changizi, Mark. 2009. **The Vision Revolution.** Dallas: BenBella Books.

Changizi, Mark A., Qiong Zhang, Hao Ye, and Shinsuke Shimojo. 2006. The structures of letters and symbols throughout human history are selected to match those found in objects in natural scenes. **American Naturalist** 167 (5): E117–39.

Charcot, J. M. 1889. **Clinical Lectures on Diseases of the Nervous System.** Vol. III, contains Lecture XI, "On a case of word-blindness," and Lecture XIII, "On a case of sudden and isolated suppression of the mental vision of signs and objects (forms and colours)." London: New Sydenham Society.

Chebat, Daniel-Robert, Constant Rainville, Ron Kupers, and Maurice Ptito. 2007. Tactile-"visual" acuity of the tongue in early blind individuals. **NeuroReport** 18: 1901–04.

Cisne, John. 2009. Stereoscopic comparison as the long-lost secret to microscopically detailed illumination like the Book of Kells. **Perception** 38 (7): 1087–1103.

Cohen, Leonardo G., Pablo Celnik, Alvaro Pascual-Leone, Brian Corwell, Lala Faiz, James Dambrosia, Manabu Honda, Norihiro Sadato, Christian Gerloff, M. Dolores Catalá, and Mark Hallett. 1997. Functional relevance of cross-modal plasticity in blind humans. **Nature** 389: 180–83.

Critchley, Macdonald. 1951. Types of visual perseveration: "paliopsia" and "illusory visual spread." **Brain** 74: 267–98.

———. 1953. **The Parietal Lobes.** New York: Hafner.

———. 1962. Dr. Samuel Johnson's aphasia. **Medical History** 6: 27–44.

Damasio, Antonio R. 2005. A mechanism for impaired fear recognition after amygdala damage. **Nature** 433 (7021): 22–23.

Damasio, Antonio R., and Hanna Damasio. 1983. The anatomic basis of pure alexia. **Neurology** 33: 1573–83.

Damasio, Antonio, Hanna Damasio, and Gary W. Van Hoesen. 1982. Prosopagnosia: Anatomic basis and behavioral mechanisms. **Neurology** 32: 331.

Darwin, Charles. 1887. **The Autobiography of Charles Darwin, 1809–1882.** Reprint, New York: W. W. Norton, 1993.

Dehaene, Stanislas. 1999. **The Number Sense.** New York: Oxford University Press.

———. 2009. **Reading in the Brain: The Science and Evolution of a Human Invention.** New York: Viking.

Déjerine, J. 1892. Contribution à l'étude anatomo-pathologique et clinique des différentes variétés de cécité verbale. **Mémoires de la Société de Biology** 4: 61–90.

Della Sala, Sergio, and Andrew W. Young. 2003. Quaglino's 1867 case of prosopagnosia. **Cortex** 39: 533–40.

Devinsky, Orrin. 2009. Delusional misidentifications and duplications. **Neurology** 72: 80–87.

Devinsky, Orrin, Lila Davachi, Cornelia Santchi, Brian T. Quinn, Bernhard P. Staresina, and Thomas Thesen. 2010. Hyperfamiliarity for faces. **Neurology** 74: 970–74.

Devinsky, Orrin, Martha J. Farah, and William B. Barr. 2008. Visual agnosia. In **Handbook of Clinical Neurology**, ed. Georg Goldenberg and Bruce Miller, vol. 88: 417–27.

Donald, Merlin. 1991. **Origins of the Modern Mind: Three Stages in the Evolution of Culture and Cognition.** Cambridge: Harvard University Press.

Duchaine, Bradley, Laura Germine, and Ken Nakayama. 2007. Family resemblance: Ten family members with prosopagnosia and within-class object agnosia. **Cognitive Neuropsychology** 24 (4): 419–30.

Duchaine, Bradley, and Ken Nakayama. 2005. Dissociations of face and object recognition in developmental prosopagnosia. **Journal of Cognitive Neuroscience** 172: 249–61.

Eling, Paul, ed. 1994. **Reader in the History of Aphasia: From Franz Gall to Norman Geschwind.** Philadelphia: John Benjamins.

Ellinwood, Everett H., Jr. 1969. Perception of faces: disorders in organic and psychopathological states. **Psychiatric Quarterly** 43 (4): 622–46.

Ellis, Hadyn D., and Melanie Florence. 1990. Bodamer's (1947) paper on prosopagnosia. **Cognitive Neuropsychology** 7 (2): 81–105.

Engel, Howard. 2005. **Memory Book.** Toronto: Penguin Canada.

———. 2007. **The Man Who Forgot How to Read.** Toronto: HarperCollins.

Etcoff, Nancy, Paul Ekman, John J. Magee, and Mark G. Frank. 2000. Lie detection and language comprehension. **Nature** 405: 139.

Farah, Martha. 2004. **Visual Agnosia,** 2nd ed. Cambridge: MIT Press/Bradford Books.

Farah, Martha, Michael J. Soso, and Richard M. Dasheiff. 1992. Visual angle of the mind's eye before and after unilateral occipital lobectomy. **Journal of Experimental Psychology: Human Perception and Performance** 18 (1): 241–46.

ffytche, D. H., R. J. Howard, M. J. Brammer, A. David, P. Woodruff, and S. Williams. 1998. The anatomy of conscious vision: an fMRI study of visual hallucinations. **Nature Neuroscience** 1 (8): 738–42.

ffytche, D. H., J. M. Lappin, and M. Philpot. 2004. Visual command hallucinations in a patient with pure alexia. **Journal of Neurology, Neurosurgery and Psychiatry** 75: 80–86.

Fleishman, John A., John D. Segall, and Frank P. Judge, Jr. 1983. Isolated transient alexia: A migrainous accompaniment. **Archives of Neurology** 40: 115–16.

Fraser, J. T. 1987. **Time, the Familiar Stranger.** Amherst: University of Massachusetts Press. (See also Foreword to the 1989 Braille edition, Stuart, FL: Triformation Braille Service.)

Freiwald, Winrich A., Doris Y. Tsao, and Margaret S. Livingstone. 2009. A face feature space in the macaque temporal lobe. **Nature Neuroscience** 12 (9): 1187–96.

Galton, Francis. 1883. **Inquiries into Human Faculty and Its Development.** London: Macmillan.

Garrido, Lucia, Nicholas Furl, Bogdan Draganski, Nikolaus Weiskopf, John Stevens, Geoffrey Chern-Yee Tan, Jon Driver, Ray J. Dolan, and Bradley Duchaine. 2009. Voxel-based morphometry reveals reduced grey matter volume in the temporal cortex of developmental prosopagnosics. **Brain** 132: 3443–55.

Gauthier, Isabel, Pawel Skudlarski, John C. Gore, and Adam W. Anderson. 2000. Expertise for cars and birds recruits brain areas involved in face recognition. **Nature Neuroscience** 3 (2): 191–97.

Gauthier, Isabel, Michael J. Tarr, and Daniel Bub, eds. 2010. **Perceptual Expertise: Bridging Brain and Behavior.** New York: Oxford University Press.

Gibson, James J. 1950. **The Perception of the Visual World.** Boston: Houghton Mifflin.

Goldberg, Elkhonon. 1989. Gradiential approach to neocortical functional organization. **Journal of Clinical and Experimental Neuropsychology** 11 (4): 489–517.

———. 2009. **The New Executive Brain: Frontal Lobes in a Complex World.** New York: Oxford University Press.

Gould, Stephen Jay. 1980. **The Panda's Thumb.** New York: W. W. Norton.

Grandin, Temple. 1996. **Thinking in Pictures: And Other Reports from My Life with Autism.** New York: Vintage.

Gregory, R. L. 1980. Perceptions as hypotheses. **Philosophical Transactions of the Royal Society, London B** 290: 181–97.

Gross, C. G. 1999. **Brain, Vision, Memory: Tales in the History of Neuroscience.** Cambridge: MIT Press/Bradford Books.

———. 2010. Making sense of printed symbols. **Science** 327: 524–25.

Gross, C. G., D. B. Bender, C. E. Rocha-Miranda. 1969. Visual receptive fields of neurons in inferotemporal cortex of the monkey. **Science** 166: 1303–6.

Gross, C. G., C. E. Rocha-Miranda, and D. B. Bender. 1972. Visual properties of neurons in inferotemporal cortex of the macaque. **Journal of Neurophysiology** 35: 96–111.

Hadamard, Jacques. 1954. **The Psychology of Invention in the Mathematical Field.** New York: Dover.

Hale, Sheila. 2007. **The Man Who Lost His Language: A Case of Aphasia.** London and Philadelphia: Jessica Kingsley.

Hamblyn, Richard. 2001. **The Invention of Clouds: How an Amateur Meteorologist Forged the Language of the Skies.** New York: Farrar, Straus and Giroux.

Harrington, Anne. 1987. **Medicine, Mind, and the Double Brain: A Study in Nineteenth-Century Thought.** Princeton: Princeton University Press.

Head, Henry. 1926. **Aphasia and Kindred Disorders of Speech.** 2 vols. Cambridge: Cambridge University Press.

Head, Henry, and Gordon Holmes. 1911. Sensory disturbances from cerebral lesions. **Brain** 34: 102–254.

Hefter, Rebecca L., Dara S. Manoach, and Jason J. S. Barton. 2005. Perception of facial expression and facial identity in subjects with social developmental disorders. **Neurology** 65: 1620–25.

Holmes, Oliver Wendell. 1861. Sun painting and sun sculpture. **Atlantic Monthly** 8: 13–29.

Hubel, David H., and Torsten N. Wiesel. 2005. **Brain and Visual Perception: The Story of a 25-Year Collaboration.** New York: Oxford University Press.

Hull, John. 1991. **Touching the Rock: An Experience of Blindness.** New York: Panthcon.

Humphreys, Glyn W., ed. 1999. **Case Studies in the Neuropsychology of Vision.** East Sussex: Psychology Press.

Judd, Tedd, Howard Gardner, and Norman Geschwind. 1983. Alexia without agraphia in a composcr. **Brain** 106: 435–57.

Julesz, Bela. 1971. **Foundations of Cyclopean Perception.** Chicago: University of Chicago Press.

Kanwisher, Nancy, Josh McDermott, and Marvin M. Chun. 1997. The fusiform face area: a module in human extrastriate cortex specialized for face perception. **Journal of Neuroscience** 17 (11): 4302–11.

Kapur, Narinder, ed. 1997. **Injured Brains of Medical Minds: Views from Within.** Oxford: Oxford University Press.

Karinthy, Frigyes. 2008. **Journey Round My Skull.** New York: NYRB Classics.

Kelly, David, Paul C. Quinn, Alan M. Slater, Kang Lee, Alan Gibson, Michael Smith, Liezhong Ge, and Olivier Pascalis. 2005. Three-month-olds, but not newborns, prefer own-race faces. **Developmental Science** 8 (6): F31–F36.

Klessinger, Nicolai, Marcin Szczerbinski, and Rosemary Varley. 2007. Algebra in a man with severe aphasia. **Neuropsychologia** 45 (8): 1642–48.

Kosslyn, Stephen Michael. 1973. Scanning visual images: Some structural implications. **Perception & Psychophysics** 14 (1): 90–94.

———. 1980. **Image and Mind.** Cambridge: Harvard University Press.

Kosslyn, Stephen M., William L. Thompson, and Giorgio Ganis. 2006. **The Case for Mental Imagery.** New York: Oxford University Press.

Lissauer, Heinrich. 1890. Ein Fall von Seelenblindheit nebst einem Beitrag zur Theorie derselben. **Archiv für Psychiatrie** 21: 222–70.

Livingstone, Margaret S., and Bevil R. Conway. 2004. Was Rembrandt stereoblind? **New England Journal of Medicine** 351 (12): 1264–65.

Luria, A. R. 1972. **The Man With a Shattered World.** New York: Basic Books.

Lusseyran, Jacques. 1998. **And There Was Light.** New York: Parabola Books.

Magee, Bryan, and Martin Milligan. 1995. **On Blindness.** New York: Oxford University Press.

Mayer, Eugene, and Bruno Rossion. 2007. Prosopagnosia. In **The Behavioral and Cognitive Neurology of Stroke,** ed. O. Godefroy and J.

Bogousslavsky, pp. 316–35. Cambridge: Cambridge University Press.

McDonald, Ian. 2006. Musical alexia with recovery: A personal account. **Brain** 129 (10): 2554–61.

McGinn, Colin. 2004. **Mindsight: Image, Dream, Meaning.** Cambridge: Harvard University Press.

Merabet, L. B., R. Hamilton, G. Schlaug, J. D. Swisher, E. T. Kiriakopoulos, N. B. Pitskel, T. Kauffman, and A. Pascual-Leone. 2008. Rapid and reversible recruitment of early visual cortex for touch. **PLoS One** Aug. 27: 3 (8): e3046.

Mesulam, M.-M. 1985. **Principles of Behavioral Neurology.** Philadelphia: F. A. Davis.

Morgan, W. Pringle. 1896. A case of congenital word blindness. **British Medical Journal** 2 (1871): 1378.

Moss, C. Scott. 1972. **Recovery with Aphasia: The Aftermath of My Stroke.** Urbana: University of Illinois Press.

Nakayama, Ken. 2001. Modularity in perception, its relation to cognition and knowledge. In **Blackwell Handbook of Perception,** ed. E. Bruce Goldstein, pp. 737–59. Malden, MA: Wiley-Blackwell.

Ostrovsky, Yuri, Aaron Andalman, and Pawan Sinha. 2006. Vision following extended congenital blindness. **Psychological Science** 17 (12): 1009–14.

Pallis, C. A. 1955. Impaired identification of faces and places with agnosia for colours. **Journal of Neurology, Neurosurgery and Psychiatry** 18: 218.

Pammer, Kristen, Peter C. Hansen, Morten L. Kringelbach, Ian Holliday, Gareth Barnes, Arjan Hillebrand, Krish D. Singh, and Piers L.

Cornelissen. 2004. Visual word recognition: the first half second. **NeuroImage** 22: 1819–25.

Pascalis, O., L. S. Scott, D. J. Kelly, R. W. Shannon, E. Nicholson, M. Coleman, and C. A. Nelson. 2005. Plasticity of face processing in infancy. **Proceedings of the National Academy of Sciences** 102 (14): 5297–5300.

Pascual-Leone, A., L. B. Merabet, D. Maguire, A. Warde, K. Alterescu, and R. Stickgold. 2004. Visual hallucinations during prolonged blindfolding in sighted subjects. **Journal of Neuroophthalmology** 24 (2): 109–13.

Petersen, S. E., P. T. Fox, M. I. Posner, M. Mintun, and M. E. Raichle. 1988. Positron emission tomographic studies of the cortical anatomy of single-word processing. **Nature** 331 (6137): 585–89.

Poe, Edgar Allan. 1846. "The Sphinx." In **Complete Stories and Poems of Edgar Allan Poe.** Reprint, New York: Doubleday, 1984.

Pomeranz, Howard D., and Simmons Lessell. 2000. Palinopsia and polyopia in the absence of drugs or cerebral disease. **Neurology** 54: 855–59.

Pons, Tim. 1996. Novel sensations in the congenitally blind. **Nature** 380: 479–80.

Prescott, William. 1843. **A History of the Conquest of Mexico: With a Preliminary View of the Ancient Mexican Civilization and the Life of Hernando Cortes.** Reprint, London: Everyman's Library, 1957.

———. 1847. **A History of the Conquest of Peru.** Reprint, London: Everyman's Library, 1934.

Ptito, Maurice, Solvej M. Moesgaard, Albert Gjedde, and Ron Kupers. 2005. Cross-modal plasticity revealed by electrotactile stimulation of the tongue in the congenitally blind. **Brain** 128 (3): 606–14.

Purves, Dale, and R. Beau Lotto. 2003. **Why We See What We Do: An Empirical Theory of Vision.** Sunderland, MA: Sinauer Associates.

Quian Quiroga, Rodrigo, Alexander Kraskov, Christof Koch, and Itzhak Fried. 2009. Explicit encoding of multimodal percepts by single neurons in the human brain. **Current Biology** 19: 1308–13.

Quian Quiroga, R., L. Reddy, G. Kreiman, C. Koch, and I. Fried. 2005. Invariant visual representation by single neurons in the human brain. **Nature** 435 (23): 1102–07.

Ramachandran, V. S. 1995. Perceptual correlates of neural plasticity in the adult human brain. In **Early Vision and Beyond,** ed. Thomas V. Papathomas, pp. 227–47. Cambridge: MIT Press/Bradford Books.

———. 2003. Foreword. In **Filling-In: From Perceptual Completion to Cortical Reorganization,** ed. Luiz Pessoa and Peter De Weerd, pp. xi–xxii. New York: Oxford University Press.

Ramachandran, V. S., and R. L. Gregory. 1991. Perceptual filling in of artificially induced scotomas in human vision. **Nature** 350 (6320): 699–702.

Renier, Laurent, and Anne G. De Volder. 2005. Cognitive and brain mechanisms in sensory substitution of vision: a contribution to the study of human perception. **Journal of Integrative Neuroscience** 4 (4): 489–503.

Rocke, Alan J. 2010. **Image and Reality: Kekulé, Kopp, and the Scientific Imagination.** Chicago: University of Chicago Press.

Romano, Paul. 2003. A case of acute loss of binocular vision and stereoscopic depth perception. **Binocular Vision & Strabismus Quarterly** 18 (1): 51–55.

Rosenfield, Israel. 1988. **The Invention of Memory.** New York: Basic Books.

Russell, R., B. Duchaine, and K. Nakayama. 2009. Super-recognizers: People with extraordinary face recognition ability. **Psychonomic Bulletin & Review** 16: 252–57.

Sacks, Oliver. 1984. **A Leg to Stand On.** New York: Summit Books.

———. 1985. **The Man Who Mistook His Wife for a Hat.** New York: Summit Books.

———. 1995. **An Anthropologist on Mars.** New York: Alfred A. Knopf.

———.1996. **The Island of the Colorblind.** New York: Alfred A. Knopf.

———. 2006. Stereo Sue. **The New Yorker** (June 19): 64–73.

———. 2008. **Musicophilia.** Rev. ed. New York: Alfred A. Knopf.

Sacks, Oliver, and Ralph M. Siegel. 2006. Seeing is believing as brain reveals its adaptability. Letter to the Editor. **Nature** 441 (7097): 1048.

Sadato, Norihiro. 2005. How the blind "see" Braille: Lessons from functional magnetic resonance imaging. **Neuroscientist** 11 (6): 577–82.

Sadato, Norihiro, Alvaro Pascual-Leone, Jordan Grafman, Vicente Ibañez, Marie-Pierre Deiber,

George Dold, and Mark Hallett. 1996. Activation of the primary visual cortex by Braille reading in blind subjects. **Nature** 380: 526–28.

Sasaki, Yuka, and Takeo Watanabe. 2004. The primary visual cortex fills in color. **Proceedings of the National Academy of Sciences of the USA** 101 (52): 18251–56.

Scribner, Charles, Jr. 1993. **In the Web of Ideas: The Education of a Publisher.** New York: Charles Scribner's Sons.

Sellers, Heather. 2007. Tell me again who you are. In **Best Creative Nonfiction,** ed. Lee Gutkind, pp. 281–319. New York: W. W. Norton.

———. 2010. **You Don't Look Like Anyone I Know.** New York: Riverhead Books.

Shallice, Tim. 1988. Lissauer on agnosia. **Cognitive Neuropsychology** 5 (2): 153–92.

Shepard, R. N., and J. Metzler. 1971. Mental rotation of three-dimensional objects. **Science** 171: 701–03.

Shimojo, Shinsuke, and Ken Nakayama. 1990. Real world occlusion constraints and binocular rivalry. **Vision Research** 30 (1): 69–80.

Shimojo, S., M. Paradiso, and I. Fujita. 2001. What visual perception tells us about mind and brain. **Proceedings of the National Academy of Sciences of the USA** 98 (22): 12340–41.

Shimojo, S., and Ladan Shams. 2001. Sensory modalities are not separate modalities: Plasticity and interactions. **Current Opinion in Neurobiology** 11: 505–09.

Shin, Yong-Wook, Myung Hyon Na, Tae Hyon Ha, Do-Hyung Kang, So-Young Yoo, and Jun

Soo Kwon. 2008. Dysfunction in configural face processing in patients with schizophrenia. **Schizophrenia Bulletin** 34 (3): 538–43.

Sugita, Yoichi. 2008. Face perception in monkeys reared with no exposure to faces. **Proceedings of the National Academy of Sciences of the USA** 105(1): 394–98.

Tanaka, Keiji. 1996. Inferotemporal cortex and object vision. **Annual Review of Neuroscience** 19: 109–39.

———. 2003. Columns for complex visual object features in the inferotemporal cortex: Clustering of cells with similar but slightly different stimulus selectivities. **Cerebral Cortex** 13 (1): 90–99.

Tarr, M. J., and I. Gauthier. 2000. FFA: A flexible fusiform area for subordinate-level visual processing automatized by expertise. **Nature Neuroscience** 3 (8): 764–69.

Temple, Christine. 1992. Developmental memory impairment: Faces and patterns. In **Mental Lives: Case Studies in Cognition,** ed. Ruth Campbell, pp. 199–215. Oxford: Blackwell.

Tenberken, Sabriye. 2003. **My Path Leads to Tibet.** New York: Arcade Publishing.

Torey, Zoltan. 1999. **The Crucible of Consciousness.** New York: Oxford University Press.

———.2003. **Out of Darkness.** New York: Picador.

Turnbull, Colin M. 1961. **The Forest People.** New York: Simon & Schuster.

West, Thomas G. 1997. **In the Mind's Eye: Visual Thinkers, Gifted People with Dyslexia and Other Learning Difficulties, Computer Images and the**

Ironies of Creativity. Amherst, NY: Prometheus Books.

Wheatstone, Charles. 1838. Contributions to the physiology of vision.—Part the first. On some remarkable, and hitherto unobserved phenomena of binocular vision. **Philosophical Transactions of the Royal Society of London** 128: 371–94.

Wigan, A. L. 1844. **The Duality of the Mind, Proved by the Structure, Functions and Diseases of the Brain.** London: Longman, Brown, Green and Longmans.

Wolf, Maryanne. 2007. **Proust and the Squid: The Story and Science of the Reading Brain.** New York: HarperCollins.

Yardley, Lucy, Lisa McDermott, Stephanie Pisarski, Brad Duchaine, and Ken Nakayama. 2008. Psychosocial consequences of developmental prosopagnosia: A problem of recognition. **Journal of Psychosomatic Research** 65: 445–51.

Zur, Dror, and Shimon Ullmann. 2003. Filling-in of retinal scotomas. **Vision Research** 43: 971–82.

INDEX

PERMISSIONS ACKNOWLEDGMENTS

Grateful acknowledgment is made to the following for permission to reprint previously published material:

Basic Books: Excerpts from **The Invention of Memory** by Israel Rosenfield, copyright © 1998 by Israel Rosenfield. Reprinted by permission of Basic Books, a member of the Perseus Books Group, administered by Copyright Clearance Center.

Pantheon Books: Excerpts from **Touching the Rock** by John Hull, copyright © 1990 by John M. Hull. Reprinted by permission of Pantheon Books, a division of Random House, Inc.

ABOUT THE AUTHOR

Oliver Sacks is a practicing physician and the author of ten books, including **Musicophilia**, **The Man who Mistook His Wife for a Hat**, and **Awakenings** (which inspired the Oscar-nominated film). He lives in New York City, where he is a professor of neurology and psychiatry at Columbia University Medical Center and the first Columbia University Artist.

www.oliversacks.com